PROBABILITY: THEORY AND EXAMPLES

Edited by
Maria Catherine C. Borres

ArclerPress

Probability: Theory and Examples

Edited by Maria Catherine C. Borres

Edition 2018

ArclerPress

Published by
Arcler Press LLC
708 3rd Avenue, 6th Floor
New York
NY 10017
United States of America
www.arclerpress.com

ISBN: 978-1-68094-587-4 (Hardback)

Printed in United States of America

Library of Congress Control Number: 2017933176

About the Editor

Maria Catherine C. Borres

Catherine is currently taking up Master of Arts in Education Major in Mathematics in Philippine Normal University – Manila. She is currently working as a Content Developer for Mathematics at the Affordable Private Education Center (APEC Schools).

Table of Contents

2 PROBABILITY INTERPRETATIONS AND BAYESIAN PROBABILITY

3 PROBABILITY DISTRIBUTIONS

4 EVENTS IN PROBABILITY THEORY

5 INDEPENDENCE PROBABILITY THEORY

6 LAW OF TOTAL PROBABILITY

7 LAW OF LARGE NUMBERS

8 VENN DIAGRAMS

List of Abbreviations

CLT	Central Limit Theorem
LLN	Law of Large Numbers

Preface

Probability theory is a fundamental pillar of modern mathematics with relations to other mathematical areas like algebra, topology, analysis, geometry or dynamical systems. The principal objects of probability theory are random variables, stochastic processes, and events: mathematical abstractions of non-deterministic events or measured quantities that may either be single occurrences or evolve over time in an apparently random fashion. Probability deals with calculating the likelihood of a given event's occurrence, which is expressed as a number between 1 and 0. The study of probability helps us figure out the likelihood of something happening. For instance, when you roll a pair of dice, you might ask how likely you are to roll a seven. In math, we call the "something happening" an "event." The probability of the occurrence of an event can be expressed as a fraction or a decimal from 0 to 1. Events that are unlikely will have a probability near 0, and events that are likely to happen have probabilities near 1. In any probability problem, it is very important to identify all the different outcomes that could occur. For instance, in the question about the dice, you must figure out all the different ways the dice could land, and all the different ways you could roll a seven. As a mathematical foundation for statistics, probability theory is essential to many human activities that involve quantitative analysis of large sets of data. Methods of probability theory also apply to descriptions of complex systems given only partial knowledge of their state, as in statistical mechanics. A great discovery of twentieth century physics was the probabilistic nature of physical phenomena at atomic scales, described in quantum mechanics. Probability theory had its start in the 17th century, when two French mathematicians, Blaise Pascal and Pierre de Fermat carried on a correspondence discussing mathematical problems dealing with games of chance. Initially, probability theory mainly considered discrete events, and its methods were mainly combinatorial. Eventually, analytical considerations compelled the incorporation of continuous variables into the theory. Contemporary applications of probability theory run the gamut of human inquiry, and include aspects of computer programming, astrophysics, music, weather prediction, and medicine. This Text contains probability basics and rules, as well as the fundamental law of total probability and Bayes' theorem.

Content Coverage

Chapters One and *Two* present facts about the probability theory with applications and probability interpretations, consistently. The probability has been used in a variety of ways since it was first applied to the mathematical study of games of chance. Does probability measure the real, physical tendency of something to occur or is it a measure of how strongly one believes it will occur, or does it draw on both these elements? In answering such questions, mathematicians interpret the probability values of probability theory.

Chapters Three and *Four* discuss about the probability distributions and probability events, respectively. Most introductions to probability theory treat discrete probability distributions and continuous probability distributions separately. In probability theory, A single outcome may be an element of many different events, and different events in an experiment are usually not equally likely, since they may include very different groups of outcomes.

Chapter Five focuses on the concept of independence extends to dealing with collections of more than two events or random variables, in which case the events are pairwise independent if each pair are independent of each other, and the events are mutually independent if each event is independent of each other combination of events..

Chapter Six emphasizes on the law of total probability relating marginal probabilities to conditional probabilities. It expresses the total probability of an outcome which can be realized through several distinct events.

Chapter Seven explicates about the law of large numbers (LLN) that describes the result of performing the same experiment a large number of times. The LLN is important because it "guarantees" stable long-term results for the averages of some random events.

Chapter Eight presents a focus on Venn diagrams that shows all possible logical relations between a finite collection of different sets. They are used to teach elementary set theory, as well as illustrate simple set relationships in probability, logic, and statistics.

CHAPTER 1

Elementary Probability

Chapter Contents

INTRODUCTION

Probability is the measure of the likelihood that an event will occur. Probability is quantified as a number between 0 and 1 (where 0 indicates impossibility and 1 indicates certainty). The higher the probability of an event, the more certain that the event will occur. A simple example is the tossing of a fair (unbiased) coin. Since the coin is unbiased, the two outcomes ("head" and "tail") are both equally probable; the probability of "head" equals the probability of "tail." Since no other outcomes are possible, the probability is 1/2 (or 50%), of either "head" or "tail". In other words, the probability of "head" is 1 out of 2 outcomes and the probability of "tail" is also 1 out of 2 outcomes, expressed as 0.5 when converted to decimal, with the above-mentioned quantification system. This type of probability is also called a priori probability.

These concepts have been given an axiomatic mathematical formalization in probability theory, which is used widely in such areas of study as mathematics, statistics, finance, gambling, science (in particular physics), artificial intelligence/machine learning, computer science, game theory, and philosophy to, for example, draw inferences about the expected frequency of events. Probability theory is also used to describe the underlying mechanics and regularities of complex systems.

INTERPRETATIONS

When dealing with experiments that are random and well-defined in a purely theoretical setting (like tossing a fair coin), probabilities can be numerically described by the number of desired outcomes divided by the total number of all outcomes. For example, tossing a fair coin twice will yield "head-head", "head-tail", "tail-head", and "tail-tail" outcomes. The probability of getting an outcome of "head-head" is 1 out of 4 outcomes or 1/4 or 0.25 (or 25%). When it comes to practical application however, there are two major competing categories of **probability interpretations**, whose adherents possess different views about the fundamental nature of probability:

- Objectivists assign numbers to describe some objective or physical state of affairs. The most popular version of objective probability is frequentist probability, which claims that the probability of a random event denotes the relative frequency of occurrence of an experiment's outcome, when repeating the experiment. This interpretation considers probability to be the relative frequency "in the long run" of outcomes. A modification of this is propensity probability, which interprets probability as the tendency of some experiment to yield a certain outcome, even if it is performed only once.
- Subjectivists assign numbers per subjective probability, i.e., as a degree of belief. The degree of belief has been interpreted as, "the price at which you would buy or sell a bet that pays 1 unit of utility if E, 0 if not E." The most popular version of subjective probability is Bayesian probability, which includes expert knowledge as well as experimental data to produce probabilities. The expert knowledge is represented by some (subjective) prior probability distribution. These data are incorporated in a likelihood function. The product of the prior and the likelihood, normalized, results in a posterior probability distribution that incorporates all the information known to date. Starting from arbitrary, subjective probabilities for a group of agents, some Bayesiansclaim that all agents will eventually have sufficiently similar assessments of probabilities, given enough evidence (see Cromwell›s rule).

ETYMOLOGY

The word probability derives from the Latin probabilitas, which can also mean "probity", a measure of the authority of a witness in a legal case in Europe, and often correlated with the witness›s nobility. In a sense, this differs much from the modern meaning of probability, which, in contrast, is a measure of the weight of empirical evidence, and is arrived at from inductive reasoning and statistical inference.

History

The scientific study of probability is a modern development of mathematics. Gambling shows that there has been an interest in quantifying the ideas of probability for millennia, but exact mathematical descriptions arose much later. There are reasons of course, for the slow development of the mathematics of probability. Whereas games of chance provided the impetus for the mathematical study of probability, fundamental issues are still obscured by the superstitions of gamblers.

Christiaan Huygens probably published the first book on probability.

According to Richard Jeffrey, "Before the middle of the seventeenth century, the term 'probable' (Latin probabilis) meant approvable, and was applied in that sense, univocally, to opinion and to action. A probable action or opinion was one such as sensible people would undertake or hold, in the circumstances." However, in legal contexts especially, 'probable' could also apply to propositions for which there was good evidence.

Gerolamo Cardano.

The sixteenth century Italian polymath Gerolamo Cardano demonstrated the efficacy of defining odds as the ratio of favourable to unfavourable outcomes (which implies that the probability of an event is given by the ratio of favourable outcomes to the total number of possible outcomes). Aside from the elementary work by Cardano, the doctrine of probabilities dates to the correspondence of Pierre de Fermat and Blaise Pascal (1654). Christiaan Huygens (1657) gave the earliest known scientific treatment of the subject. Jakob Bernoulli›s Ars Conjectandi (posthumous, 1713) and Abraham de Moivre›s Doctrine of Chances (1718) treated the subject as a branch of mathematics. See Ian Hacking›s The Emergence of Probability and James Franklin›s The Science of Conjecture for histories of the early development of the very concept of mathematical probability.

The theory of errors may be traced back to Roger Cotes›s Opera Miscellanea (posthumous, 1722), but a memoir prepared by Thomas Simpson in 1755 (printed 1756) first applied the theory to the discussion of errors of observation. The reprint (1757) of this memoir lays down the axioms that positive and negative errors are equally probable, and that certain assignable limits define the range of all errors. Simpson also discusses continuous errors and describes a probability curve.

The first two laws of error that were proposed both originated with Pierre-Simon Laplace. The first law was published in 1774 and stated that the frequency of an error could be expressed as an exponential function of the numerical magnitude of the error, disregarding sign. The second law of error was proposed in 1778 by Laplace and stated that the frequency of the error is an exponential function of the square of the error. The second law of error is called the normal distribution or the Gauss law. "It is difficult historically to attribute that law to Gauss, who in spite of his well-known precocity had probably not made this discovery before he was two years old. "

Daniel Bernoulli (1778) introduced the principle of the maximum product of the probabilities of a system of concurrent errors.

Adrien-Marie Legendre (1805) developed the method of least squares, and introduced it in his Nouvelles méthodes pour la détermination des orbites des comètes (New Methods for Determining the Orbits of Comets). In ignorance of Legendre's contribution, an Irish-American writer, Robert Adrain, editor of "The Analyst" (1808), first deduced the law of facility of error,

$$\phi(x) = ce^{-h^2x^2}$$

where h is a constant depending on precision of observation, and c is a scale factor ensuring that the area under the curve equals 1. He gave two proofs, the second being essentially the same as John Herschel›s (1850). Gauss gave the first proof that seems to have been known in Europe (the third after Adrain›s) in 1809. Further proofs were given by Laplace (1810, 1812), Gauss (1823), James Ivory (1825, 1826), Hagen (1837), Friedrich Bessel (1838), W. F. Donkin (1844, 1856), and Morgan Crofton (1870). Other contributors were Ellis (1844), De Morgan (1864), Glaisher (1872), and Giovanni Schiaparelli (1875). Peters›s (1856) formula for r, the probable error of a single observation, is well known.

Carl Friedrich Gauss.

In the nineteenth century authors on the general theory included Laplace, Sylvestre Lacroix (1816), Littrow (1833), Adolphe Quetelet (1853), Richard Dedekind (1860), Helmert (1872), Hermann Laurent (1873), Liagre, Didion, and Karl Pearson. Augustus De Morgan and George Boole improved the exposition of the theory.

Andrey Markov introduced the notion of Markov chains (1906), which played an important role in stochastic processes theory and its applications. The modern theory of probability based on the measure theory was developed by Andrey Kolmogorov (1931).

On the geometric side (see integral geometry) contributors to The Educational Times were influential (Miller, Crofton, McColl, Wolstenholme, Watson, and Artemas Martin).

THEORY

Like other theories, the theory of probability is a representation of probabilistic concepts in formal terms—that is, in terms that can be considered separately from their meaning. These formal terms are manipulated by the rules of mathematics and logic, and any results are interpreted or translated back into the problem domain.

There have been at least two successful attempts to formalize probability, namely the Kolmogorov formulation and the Cox formulation. In Kolmogorov's formulation (see probability space), sets are interpreted as events and probability itself as a measure on a class of sets. In Cox›s theorem, probability is taken as a primitive (that is, not further analyzed) and the emphasis is on constructing a consistent assignment of probability values to propositions. In both cases, the laws of probability are the same, except for technical details.

There are other methods for quantifying uncertainty, such as the Dempster–Shafer theory or possibility theory, but those are essentially different and not compatible with the laws of probability as usually understood.

APPLICATIONS

Probability theory is applied in everyday life in risk assessment and modeling. The insurance industry and markets use actuarial science to determine pricing and make trading decisions. Governments apply probabilistic methods in environmental regulation, entitlement analysis (Reliability theory of aging and longevity), and financial regulation.

A good example of the use of probability theory in equity trading is the effect of the perceived probability of any widespread Middle East conflict on oil prices, which have ripple effects in the economy as a whole. An assessment by a commodity trader that a war is more likely can send that commodity›s prices up or down, and signals other traders of that opinion. Accordingly, the probabilities are neither assessed independently nor necessarily very rationally. The theory of behavioral finance emerged to describe the effect of such groupthink on pricing, on policy, and on peace and conflict.

In addition to financial assessment, probability can be used to analyze trends in biology (e.g. disease spread) as well as ecology (e.g. biological Punnett squares). As with finance, risk assessment can be used as a statistical tool to calculate the likelihood of undesirable events occurring and can assist with implementing protocols to avoid encountering such circumstances. Probability is used to design games of chance so that casinos can make a guaranteed profit, yet provide payouts to players that are frequent enough to encourage continued play.

The discovery of rigorous methods to assess and combine probability assessments has changed society. It is important for most citizens to understand how probability assessments are made, and how they contribute to decisions.

Another significant application of probability theory in everyday life is reliability. Many consumer products, such as automobiles and consumer electronics, use reliability theory in product design to reduce the probability of failure. Failure probability may influence a manufacturer's decisions on a product's warranty.

The cache language model and other statistical language models that are used in natural language processing are also examples of applications of probability theory.

MATHEMATICAL TREATMENT

Consider an experiment that can produce a number of results. The collection of all possible results is called the sample space of the experiment. The power set of the sample space is formed by considering all different collections of possible results. For example, rolling a dice can produce six possible results. One collection of possible results gives an odd number on the dice. Thus, the subset $\{1,3,5\}$ is an element of the power set of the sample space of dice rolls. These collections are called "events." In this case, {1,3,5} is the event that the dice falls on some odd number. If the results that actually occur fall in a given event, the event is said to have occurred.

A probability is a way of assigning every event a value between zero and one, with the requirement that the event made up of all possible results (in our example, the event {1,2,3,4,5,6}) is assigned a value of one. To qualify as a probability, the assignment of values must satisfy the requirement that if you look at a collection of mutually exclusive events (events with no common results, e.g., the events {1,6}, {3}, and {2,4} are all mutually exclusive), the probability that at least one of the events will occur is given by the sum of the probabilities of all the individual events.

The probability of an event A is written as $P(A)$, $p(A)$, or $\Pr(A)$. This mathematical definition of probability can extend to infinite sample spaces, and even uncountable sample spaces, using the concept of a measure.

The opposite or complement of an event A is the event [not A] (that is, the event of A not occurring), often denoted as $\overline{A}, A^C, \neg A$, or $\sim A$; its probability is given by P(not A) = 1 − P(A). As an example, the chance of not rolling a six on a six-sided die is 1 – (chance of rolling a six) $= 1 - \frac{1}{6} = \frac{5}{6}$. See Complementary event for a more complete treatment.

If two events A and B occur on a single performance of an experiment, this is called the intersection or joint probability of A and B, denoted as $P(A \cap B)$.

Independent Events

If two events, A and B are independent then the joint probability is

$$P(A \text{ and } B) = P(A \cap B) = P(A)P(B),$$

for example, if two coins are flipped the chance of both being heads is $\frac{1}{2} \times \frac{1}{2} = \frac{1}{4}$.

Mutually Exclusive Events

If either event A or event B occurs on a single performance of an experiment this is called the union of the events A and B denoted as as $P(A \cup B)$. If two events are mutually exclusive then the probability of either occurring is

$$P(A \text{ or } B) = P(A \cup B) = P(A) + P(B).$$

For example, the chance of rolling a 1 or 2 on a six-sided die is $P(1 \text{ or } 2) = P(1) + P(2) = \frac{1}{6} + \frac{1}{6} = \frac{1}{3}$.

Not mutually Exclusive Events

If the events are not mutually exclusive then

$$P(A \text{ or } B) = P(A) + P(B) - P(A \text{ and } B).$$

For example, when drawing a single card at random from a regular deck of cards, the chance of getting a heart or a face card (J,Q,K) (or one that is both) is $\frac{13}{52} + \frac{12}{52} - \frac{3}{52} = \frac{11}{26}$, because of the 52 cards of a deck 13 are hearts, 12 are face cards, and 3 are both: here the possibilities included in the "3 that are both" are included in each of the "13 hearts" and the "12 face cards" but should only be counted once.

Conditional Probability

Conditional probability is the probability of some event A, given the occurrence of some other event B. Conditional probability is written $P(A \mid B)$, and is read "the probability of A, given B". It is defined by

$$P(A \mid B) = \frac{P(A \cap B)}{P(B)}.$$

If $P(B) = 0$ then $P(A \mid B)$ is formally undefined by this expression. However, it is possible to define a conditional probability for some zero-probability events using a σ-algebra of such events (such as those arising from a continuous random variable).

For example, in a bag of 2 red balls and 2 blue balls (4 balls in total), the probability of taking a red ball is $1/2$; however, when taking a second ball, the probability of it being either a red ball or a blue ball depends on the ball previously taken, such as, if a red ball was taken, the probability of picking a red ball again would be $1/3$ since only 1 red and 2 blue balls would have been remaining.

Inverse Probability

In probability theory and applications, Bayes' rule relates the odds of event A_1 to event A_2 before (prior to) and after (posterior to) conditioning on another event B. The odds on A_1 to event A_2 is simply the ratio of the probabilities of the two events. When arbitrarily many events A are of interest, not just two, the rule can be rephrased as posterior is proportional to prior times likelihood, $P(A|B) \propto P(A)P(B|A)$ where the proportionality symbol means that the left hand side is proportional to (i.e., equals a constant times) the right hand side as A varies, for fixed or given B (Lee, 2012; Bertsch McGrayne, 2012). In this form it goes back to Laplace (1774) and to Cournot (1843); see Fienberg (2005). See Inverse probability and Bayes' rule.

Summary of Probabilities

Summary of probabilities

Event	Probability
A	$P(A) \in [0,1]$
not A	$P(A^c) = 1 - P(A)$
A or B	$P(A \cup B) = P(A) + P(B) - P(A \cap B)$ $P(A \cup B) = P(A) + P(B)$ **if A and B are mutually exclusive**
A and B	$P(A \cap B) = P(A\|B)P(B) = P(B\|A)P(A)$ $P(A \cap B) = P(A)P(B)$ **if A and B are independent**
A given B	$P(A \mid B) = \frac{P(A \cap B)}{P(B)} = \frac{P(B\|A)P(A)}{P(B)}$

RELATION TO RANDOMNESS

In a deterministic universe, based on Newtonian concepts, there would be no probability if all conditions were known (Laplace›s demon), (but there are situations in which sensitivity to initial conditions exceeds our ability to measure them, i.e. know them). In the case of a roulette wheel, if the force of the hand and the period of that force are known, the number on which the ball will stop would be a certainty (though as a practical matter, this would likely be true only of a roulette wheel that had not been exactly levelled — as Thomas A. Bass' Newtonian Casino revealed). Of course, this also assumes knowledge of inertia and friction of the wheel, weight, smoothness and roundness of the ball, variations in hand speed during the turning and so forth. A probabilistic description can thus be more useful than Newtonian mechanics for analyzing the pattern of outcomes of repeated rolls of a roulette wheel. Physicists face the same situation in kinetic theory of gases, where the system, while deterministic in principle, is so complex (with the number of molecules typically the order of magnitude of Avogadro constant $6.02 \cdot 10^{23}$) that only a statistical description of its properties is feasible.

Probability theory is required to describe quantum phenomena. A revolutionary discovery of early 20th century physics was the random character of all physical processes that occur at sub-atomic scales and are governed by the laws of quantum mechanics. The objective wave function evolves deterministically but, according to the Copenhagen interpretation, it deals with probabilities of observing, the outcome being explained by a wave function collapse when an observation is made. However, the loss of determinism for the sake of instrumentalism did not meet with universal approval. Albert Einstein famously remarked in a letter to Max Born: "I am convinced that God does not play dice ". Like Einstein, Erwin Schrödinger, who discovered the wave function, believed quantum mechanics is a statistical approximation of an underlying deterministic reality. In some modern interpretations of the statistical mechanics of measurement, quantum decoherence is invoked to account for the appearance of subjectively probabilistic experimental outcomes.

PROBABILITY AXIOMS

In Kolmogorov's probability theory, the probability P of some event E, denoted {\ displaystyle P(E)}, is usually defined such that P satisfies the **Kolmogorov axioms**, named after the Russian mathematician Andrey Kolmogorov, which are described below.

These assumptions can be summarised as follows: Let (Ω, F, P) be a measure space with P(Ω) = 1. Then (Ω, F, P) is a probability space, with sample space Ω, event space F and probability measure P.

An alternative approach to formalising probability, favoured by some Bayesians, is given by Cox's theorem.

Axioms

First Axiom

The probability of an event is a non-negative real number:

$$P(E) \in \mathbb{R}, P(E) \geq 0 \qquad \forall E \in F$$

where F is the event space. In particular, $P(E)$ is always finite, in contrast with more general measure theory. Theories which assign negative probability relax the first axiom.

Second Axiom

This is the assumption of unit measure: that the probability that at least one of the elementary events in the entire sample space will occur is 1.

$$P(\Omega) = 1.$$

Third Axiom

This is the assumption of σ-additivity:

Any countable sequence of disjoint sets (synonymous with mutually exclusive events) $E_1, E_2, \ldots$ satisfies

$$P\left(\bigcup_{i=1}^{\infty} E_i\right) = \sum_{i=1}^{\infty} P(E_i).$$

Some authors consider merely finitely additive probability spaces, in which case one just needs an algebra of sets, rather than a σ-algebra. Quasiprobability distributions in general relax the third axiom.

Consequences

From the Kolmogorov axioms, one can deduce other useful rules for calculating probabilities.

The probability of the Empty Set

$P(\varnothing) = 0.$

In some cases, $\varnothing$ is not the only event with probability 0.

Monotonicity

$$\text{if} \quad A \subseteq B \quad \text{then} \quad P(A) \leq P(B).$$

If A is a subset of, or equal to B, then the probability of A is less than, or equal to the probability of B.

The numeric bound

It immediately follows from the monotonicity property that

$0 \leq P(E) \leq 1 \qquad \forall E \in F.$

Proofs

The proofs of these properties are both interesting and insightful. They illustrate the power of the third axiom, and its interaction with the remaining two axioms. When studying axiomatic probability theory, many deep consequences follow from merely these three axioms. In order to verify the monotonicity property, we set $E_1 = A$ and $E_2 = B \setminus A$, , where $A \subseteq B$ and $E_i = \varnothing$ for $i \geq 3$. . It is easy to see that the sets E_i are pairwise disjoint and $E_1 \cup E_2 \cup \cdots = B$. Hence, we obtain from the third axiom that

$$P(A) + P(B \setminus A) + \sum_{i=3}^{\infty} P(\varnothing) = P(B).$$

Since the left-hand side of this equation is a series of non-negative numbers, and that it converges to $P(B)$ which is finite, we obtain both $P(A) \leq P(B)$ and $P(\varnothing) = 0$. The second part of the statement is seen by contradiction: if $P(\varnothing) = a$ then the left hand side is not less than infinity

$$\sum_{i=3}^{\infty} P(E_i) = \sum_{i=3}^{\infty} P(\varnothing) = \sum_{i=3}^{\infty} a = \begin{cases} 0 & \text{if } a = 0, \\ \infty & \text{if } a > 0. \end{cases}$$

If $a > 0$ then we obtain a contradiction, because the sum does not exceed $P(B)$ which is finite. Thus, $a = 0$. We have shown as a byproduct of the proof of monotonicity that $P(\varnothing) = 0$.

Further Consequences

Another important property is:

$P(A \cup B) = P(A) + P(B) - P(A \cap B).$

This is called the addition law of probability, or the sum rule. That is, the probability that A or B will happen is the sum of the probabilities that A will happen and that B will happen, minus the probability that both A and B will happen. The proof of this is as follows:

$P(A \cup B) = P(A) + P(B \setminus (A \cap B))$ (by Axiom 3)

now, $P(B) = P(B \setminus (A \cap B)) + P(A \cap B)$. Eliminating $P(B \setminus (A \cap B))$ from both equations gives us the desired result.

An extension of the addition law to any number of sets is the principle of inclusion-exclusion.

Setting B to the complement A^c of A in the addition law gives

$P(A^c) = P(\Omega \setminus A) = 1 - P(A)$

That is, the probability that any event will not happen (or the event's complement) is 1 minus the probability that it will.

Simple example: coin toss

Consider a single coin-toss, and assume that the coin will either land heads (H) or tails (T) (but not both). No assumption is made as to whether the coin is fair.

We may define:

$$\Omega = \{H, T\}$$

$$F = \{\varnothing, \{H\}, \{T\}, \{H, T\}\}$$

Kolmogorov's axioms imply that:

$$P(\varnothing) = 0$$

The probability of *neither* heads *nor* tails, is 0.

$$P(\{H, T\}) = 1$$

The probability of *either* heads *or* tails, is 1.

$$P(\{H\}) + P(\{T\}) = 1$$

The sum of the probability of heads and the probability of tails, is 1.

BOOLE'S INEQUALITY

In probability theory, **Boole's inequality**, also known as the **union bound**, says that for any finite or countable set of events, the probability that at least one of the events happens is no greater than the sum of the probabilities of the individual events. Boole›s inequality is named after George Boole.

Formally, for a countable set of events $A_1, A_2, A_3, \ldots$, we have

$$\mathbb{P}\left(\bigcup_i A_i\right) \leq \sum_i \mathbb{P}(A_i).$$

In measure-theoretic terms, Boole's inequality follows from the fact that a measure (and certainly any probability measure) is σ-sub-additive.

Proof

Proof Using Induction

Boole's inequality may be proved for finite collections of events using the method of induction.

For the $n = 1$ case, it follows that

$$\mathbb{P}(A_1) \leq \mathbb{P}(A_1).$$

For the case n, we have

$$\mathbb{P}\left(\bigcup_{i=1}^{n} A_i\right) \leq \sum_{i=1}^{n} \mathbb{P}(A_i).$$

Since $\mathbb{P}(A \cup B) = \mathbb{P}(A) + \mathbb{P}(B) - \mathbb{P}(A \cap B)$, and because the union operation is associative, we have

$$\mathbb{P}\left(\bigcup_{i=1}^{n+1} A_i\right) = \mathbb{P}\left(\bigcup_{i=1}^{n} A_i\right) + \mathbb{P}(A_{n+1}) - \mathbb{P}\left(\bigcup_{i=1}^{n} A_i \cap A_{n+1}\right).$$

Since

$$\mathbb{P}\left(\bigcup_{i=1}^{n} A_i \cap A_{n+1}\right) \geq 0,$$

by the first axiom of probability, we have

$$\mathbb{P}\left(\bigcup_{i=1}^{n+1} A_i\right) \leq \mathbb{P}\left(\bigcup_{i=1}^{n} A_i\right) + \mathbb{P}(A_{n+1}),$$

and therefore

$$\mathbb{P}\left(\bigcup_{i=1}^{n+1} A_i\right) \leq \sum_{i=1}^{n} \mathbb{P}(A_i) + \mathbb{P}(A_{n+1}) = \sum_{i=1}^{n+1} \mathbb{P}(A_i).$$

Proof without Using Induction

For any events in $A_1 \ldots A_i$ in our probability space we have

$$\mathbb{P}(\bigcup_i A_i) \leq \sum_i \mathbb{P}(A_i)$$

One of the axioms of a probability space is that if $B_1 \dots B_i$ are disjoint subsets of the probability space then

$$\mathbb{P}(\bigcup_i B_i) = \sum_i \mathbb{P}(B_i)$$

this is called countable additivity.

If $B \subset A$ then $\mathbb{P}(B) \leq \mathbb{P}(A)$

Indeed, from the axioms of a probability distribution,

$$\mathbb{P}(A) = \mathbb{P}(B) + \mathbb{P}(A - B)$$

Note that both terms on the right are nonnegative.

Now we have to modify the sets A_i, so they become disjoint.

$$B_i = A_i - \bigcup_{j=1}^{i-1} A_j$$

So if $B_i \subset A_i$, then we know

$$\bigcup_{i=1}^{\infty} B_i = \bigcup_{i=1}^{\infty} A_i$$

Therefore can we make following equation

$$\mathbb{P}(\bigcup_i A_i) = \mathbb{P}(\bigcup_i B_i) = \sum_i \mathbb{P}(B_i) \leq \sum_i \mathbb{P}(A_i)$$

Bonferroni Inequalities

Boole's inequality may be generalised to find upper and lower bounds on the probability of finite unions of events. These bounds are known as **Bonferroni inequalities**, after Carlo Emilio Bonferroni, see Bonferroni (1936).

Define

$$S_1 := \sum_{i=1}^{n} \mathbb{P}(A_i),$$

and

$$S_2 := \sum_{1 \leq i < j \leq n} \mathbb{P}(A_i \cap A_j),$$

as well as

$$S_k := \sum_{1 \leq i_1 < \cdots < i_k \leq n} \mathbb{P}(A_{i_1} \cap \cdots \cap A_{i_k})$$

for all integers k in {3, ..., n}.

Then, for odd k in {1, ..., n},

$$\mathbb{P}\left(\bigcup_{i=1}^{n} A_i\right) \leq \sum_{j=1}^{k} (-1)^{j-1} S_j,$$

and for even k in {2, ..., n},

$$\mathbb{P}\left(\bigcup_{i=1}^{n} A_i\right) \geq \sum_{j=1}^{k} (-1)^{j-1} S_j.$$

Boole's inequality is recovered by setting k = 1. When k = n, then equality holds and the resulting identity is the inclusion–exclusion principle.

REFERENCES

1. "Probability". Webster›s Revised Unabridged Dictionary. G & C Merriam, 1913
2. Strictly speaking, a value of 0 indicates impossibility only when a sample space is finite. In an infinite sample space, a probability of 0 may indicate infinitesimal likelihood, but not impossibility. For example, the probability of any specific integer, say 5, being chosen completely at random from the entire set of integers is 0. However, it is not impossible for 5 to be chosen, just extremely unlikely.
3. "Kendall›s Advanced Theory of Statistics, Volume 1: Distribution Theory ", Alan Stuart and Keith Ord, 6th Ed, (2009), ISBN 9780534243128
4. William Feller, "An Introduction to Probability Theory and Its Applications", (Vol 1), 3rd Ed, (1968), Wiley, ISBN 0-471-25708-7
5. Probability Theory The Britannica website
6. Hacking, Ian (1965). The Logic of Statistical Inference. Cambridge University Press. ISBN 0-521-05165-7.
7. Finetti, Bruno de (1970). "Logical foundations and measurement of subjective probability ". Acta Psychologica. **34**: 129–145. doi:10.1016/0001-6918(70)90012-0.
8. Hájek, Alan. "Interpretations of Probability ". The Stanford Encyclopedia of Philosophy (Winter 2012 Edition), Edward N. Zalta (ed.). Retrieved 22 April 2013.
9. Hogg, Robert V.; Craig, Allen; McKean, Joseph W. (2004). Introduction to Mathematical Statistics (6th ed.). Upper Saddle River: Pearson. ISBN 0-13-008507-3.
10. Hacking, I. (2006) The Emergence of Probability: A Philosophical Study of Early Ideas about Probability, Induction and Statistical Inference, Cambridge University Press, ISBN 978-0-521-68557-3
11. Freund, John. (1973) Introduction to Probability. Dickenson ISBN 978-0822100782 (p. 1)

12. Jeffrey, R.C., Probability and the Art of Judgment, Cambridge University Press. (1992). pp. 54-55 . ISBN 0-521-39459-7
13. Franklin, J. (2001) The Science of Conjecture: Evidence and Probability Before Pascal, Johns Hopkins University Press. (pp. 22, 113, 127)
14. Some laws and problems in classical probability and how Cardano anticipated them Gorrochum, P. Chance magazine 2012
15. Abrams, William, A Brief History of Probability, Second Moment, retrieved 2008-05-23
16. Ivancevic, Vladimir G.; Ivancevic, Tijana T. (2008). Quantum leap : from Dirac and Feynman, across the universe, to human body and mind. Singapore ; Hackensack, NJ: World Scientific. p. 16. ISBN 978-981-281-927-7.
17. Franklin, James (2001). The Science of Conjecture: Evidence and Probability Before Pascal. Johns Hopkins University Press. ISBN 0801865697.
18. Wilson EB (1923) "First and second laws of error ". Journal of the American Statistical Association, 18, 143
19. Seneta, Eugene William. " "Adrien-Marie Legendre " (version 9) ". StatProb: The Encyclopedia Sponsored by Statistics and Probability Societies. Retrieved 27 January 2016.
20. http://www.statslab.cam.ac.uk/~rrw1/markov/M.pdf
21. Vitanyi, Paul M.B. (1988). "Andrei Nikolaevich Kolmogorov ". CWI Quarterly (1): 3–18. Retrieved 27 January 2016.
22. Singh, Laurie (2010) "Whither Efficient Markets? Efficient Market Theory and Behavioral Finance". The Finance Professionals' Post, 2010.
23. Gao, J.Z.; Fong, D.; Liu, X. (April 2011). "Mathematical analyses of casino rebate systems for VIP gambling ". International Gambling Studies. **11** (1): 93–106. doi:10.1080/14459795.2011.552575.
24. Gorman, Michael (2011) «Management Insights «. Management Science
25. Ross, Sheldon. A First course in Probability, 8th Edition. Page 26-27.
26. Olofsson (2005) Page 8.
27. Olofsson (2005), page 9
28. Olofsson (2005) page 35.
29. Olofsson (2005) page 29.
30. Burgi, Mark (2010) "Interpretations of Negative Probabilities ", p. 1. arXiv:1008.1287v1
31. Jedenfalls bin ich überzeugt, daß der Alte nicht würfelt. Letter to Max Born, 4 December 1926, in: Einstein/Born Briefwechsel 1916-1955.
32. Moore, W.J. (1992). Schrödinger: Life and Thought. Cambridge University Press. p. 479. ISBN 0-521-43767-9.

CHAPTER 2

Probability Interpretations and Bayesian Probability

Chapter Contents

PROBABILITY INTERPRETATIONS

The word **probability** has been used in a variety of ways since it was first applied to the mathematical study of games of chance. Does probability measure the real, physical tendency of something to occur or is it a measure of how strongly one believes it will occur, or does it draw on both these elements? In answering such questions, mathematicians interpret the probability values of probability theory.

There are two broad categories of **probability interpretations** which can be called "physical" and "evidential" probabilities. Physical probabilities, which are also called objective or frequency probabilities, are associated with random physical systems such as roulette wheels, rolling dice and radioactive atoms. In such systems, a given type of event (such as a die yielding a six) tends to occur at a persistent rate, or "relative frequency", in a long run of trials. Physical probabilities either explain, or are invoked to explain, these stable frequencies. The two main kinds of theory of physical probability are frequentist accounts (such as those of Venn, Reichenbach and von Mises) and propensity accounts (such as those of Popper, Miller, Giere and Fetzer).

Evidential probability, also called Bayesian probability, can be assigned to any statement whatsoever, even when no random process is involved, as a way to represent its subjective plausibility, or the degree to which the statement is supported by the available evidence. On most accounts, evidential probabilities are considered to be degrees of belief, defined in terms of dispositions to gamble at certain odds. The four main evidential interpretations are the classical (e.g. Laplace`s)interpretation, the subjective interpretation (de Finetti and Savage), the epistemic or inductive interpretation (Ramsey,Cox) and the logical interpretation (Keynes and Carnap). There are also evidential interpretations of probability covering groups, which are often labelled as ‹intersubjective` (proposed by Gillies and Rowbottom).

Some interpretations of probability are associated with approaches to statistical inference, including theories of estimation and hypothesis testing. The physical interpretation, for example, is taken by followers of "frequentist" statistical methods, such as Ronald Fisher, Jerzy Neyman and Egon Pearson. Statisticians of the opposing Bayesian school typically accept the existence and importance of physical probabilities, but also consider the calculation of evidential probabilities to be both valid and necessary in statistics. This article, however, focuses on the interpretations of probability rather than theories of statistical inference.

The terminology of this topic is rather confusing, in part because probabilities are studied within a variety of academic fields. The word "frequentist" is especially tricky. To philosophers it refers to a particular theory of physical probability, one that has more or less been abandoned. To scientists, on the other hand, "frequentist probability" is just another name for physical (or objective) probability. Those who promote Bayesian inference view "frequentist statistics" as an approach to statistical inference that recognises only physical probabilities. Also the word "objective", as applied to probability, sometimes means exactly what "physical" means here, but is also used of evidential probabilities that are fixed by rational constraints, such as logical and epistemic probabilities.

It is unanimously agreed that statistics depends somehow on probability. But, as to what probability is and how it is connected with statistics, there has seldom been such complete disagreement and breakdown of communication since the Tower of Babel. Doubtless, much of the disagreement is merely terminological and would disappear under sufficiently sharp analysis.

Philosophy

The **philosophy of probability** presents problems chiefly in matters of epistemology and the uneasy interface between mathematical concepts and ordinary language as it is used by non-mathematicians. Probability theory is an established field of study in mathematics. It has its origins in correspondence discussing the mathematics of games of chance between Blaise Pascal and Pierre de Fermat in the seventeenth century, and was formalized and rendered axiomatic as a distinct branch of mathematics by Andrey Kolmogorov in the twentieth century. In axiomatic form, mathematical statements about probability theory carry the same sort of epistemological confidence within the philosophy of mathematics as are shared by other mathematical statements.

The mathematical analysis originated in observations of the behaviour of game equipment such as playing cards and dice, which are designed specifically to introduce random and equalized elements; in mathematical terms, they are subjects of indifference. This is not the only way probabilistic statements are used in ordinary human language: when people say that "*it will probably rain*", they typically do not mean that the outcome of rain versus not-rain is a random factor that the odds currently favor; instead, such statements are perhaps better understood as qualifying their expectation of rain with a degree of confidence. Likewise, when it is written that "the most probable explanation" of the name of Ludlow, Massachusetts "is that it was named after Roger Ludlow", what is meant here is not that Roger Ludlow is favored by a random factor, but rather that this is the most plausible explanation of the evidence, which admits other, less likely explanations.

Thomas Bayes attempted to provide a logic that could handle varying degrees of confidence; as such, Bayesian probability is an attempt to recast the representation of probabilistic statements as an expression of the degree of confidence by which the beliefs they express are held.

Though probability initially had somewhat mundane motivations, its modern influence and use is widespread ranging from evidence-based medicine, through Six sigma, all the way to the Probabilistically checkable proof and the String theory landscape.

A summary of some interpretations of probability				
	Classical	**Frequentist**	**Subjective**	**Propensity**
Main hypothesis	Principle of indifference	Frequency of occurrence	Degree of belief	Degree of causal connection
Conceptual basis	Hypothetical symmetry	Past data and reference class	Knowledge and intuition	Present state of system
Conceptual approach	Conjectural	Empirical	Subjective	Metaphysical
Single case possible	Yes	No	Yes	Yes
Precise	Yes	No	No	Yes
Problems	Ambiguity in principle of indifference	Circular definition	Reference class problem	Disputed concept

CLASSICAL DEFINITION

The first attempt at mathematical rigour in the field of probability, championed by Pierre-Simon Laplace, is now known as the **classical definition**. Developed from studies of games of chance (such as rolling dice) it states that probability is shared equally between all the possible outcomes, provided these outcomes can be deemed equally likely. (3.1)

The theory of chance consists in reducing all the events of the same kind to a certain number of cases equally possible, that is to say, to such as we may be equally undecided about in regard to their existence, and in determining the number of cases favorable to the event whose probability is sought. The ratio of this number to that of all the cases possible is the measure of this probability, which is thus simply a fraction whose numerator is the number of favorable cases and whose denominator is the number of all the cases possible.

—Pierre-Simon Laplace, A Philosophical Essay on Probabilities

The classical definition of probability works well for situations with only a finite number of equally-likely outcomes.

This can be represented mathematically as follows: If a random experiment can result in N mutually exclusive and equally likely outcomes and if N_A of these outcomes result in the occurrence of the event A, the **probability of A** is defined by

$$P(A) = \frac{N_A}{N}.$$

There are two clear limitations to the classical definition. Firstly, it is applicable only to situations in which there is only a 'finite' number of possible outcomes. But some important random experiments, such as tossing a coin until it rises heads, give rise to an infinite set of outcomes. And secondly, you need to determine in advance that all the possible outcomes are equally likely without relying on the notion of probability to avoid circularity—for instance, by symmetry considerations.

FREQUENTISM

For frequentists, the probability of the ball landing in any pocket can be determined only by repeated trials in which the observed result converges to the underlying probability *in the long run*.

Frequentists posit that the probability of an event is its relative frequency over time,(3.4) i.e., its relative frequency of occurrence after repeating a process a large number of times under similar conditions. This is also known as aleatory probability. The events are assumed to be governed by some random physical phenomena, which are either phenomena that are predictable, in principle, with sufficient information (see Determinism); or phenomena which are essentially unpredictable. Examples of the first kind include tossing dice or spinning a roulette wheel; an example of the second kind is radioactive decay. In the case of tossing a fair coin, frequentists say that the probability of getting a heads is 1/2, not because there are two equally likely outcomes but because repeated series of large numbers of trials demonstrate that the empirical frequency converges to the limit 1/2 as the number of trials goes to infinity.

If we denote by n_a the number of occurrences of an event $\mathcal{A}$ in n trials, then if $\lim_{n \to +\infty} \frac{n_a}{n} = p$ we say that $P(\mathcal{A}) = p$.

The frequentist view has its own problems. It is of course impossible to actually perform an infinity of repetitions of a random experiment to determine the probability of an event. But if only a finite number of repetitions of the process are performed, different relative frequencies will appear in different series of trials. If these relative frequencies are to define the probability, the probability will be slightly different every time it is measured. But the real probability should be the same every time. If we acknowledge the fact that we only can measure a probability with some error of measurement attached, we still get into problems as the error of measurement can only be expressed as a probability, the very concept we are trying to define. This renders even the frequency definition circular; see for example "What is the Chance of an Earthquake?"

LOGICAL, EPISTEMIC, AND INDUCTIVE PROBABILITY

It is widely recognized that the term "probability" is sometimes used in contexts where it has nothing to do with physical randomness. Consider, for example, the claim that the extinction of the dinosaurs was **probably** caused by a large meteorite hitting the earth. Statements such as "Hypothesis H is probably true" have been interpreted to mean that the (presently available) empirical evidence (E, say) supports H to a high degree. This degree of support of H by E has been called the **logical** probability of H given E, or the **epistemic probability** of H given E, or the **inductive probability** of H given E.

The differences between these interpretations are rather small, and may seem inconsequential. One of the main points of disagreement lies in the relation between probability and belief. Logical probabilities are conceived (for example in Keynes` Treatise on Probability) to be objective, logical relations between propositions (or sentences), and hence not to depend in any way upon belief. They are degrees of (partial) entailment, or degrees of logical consequence, not degrees of belief. (They do, nevertheless, dictate proper degrees of belief, as is discussed below.) Frank P. Ramsey, on the other hand, was skeptical about the existence of such objective logical relations and argued that (evidential) probability is "the logic of partial belief".(p 157) In other words, Ramsey held that epistemic probabilities simply *are* degrees of rational belief, rather than being logical relations that merely *constrain* degrees of rational belief.

Another point of disagreement concerns the *uniqueness* of evidential probability, relative to a given state of knowledge. Rudolf Carnap held, for example, that logical principles always determine a unique logical probability for any statement, relative to any body of evidence. Ramsey, by contrast, thought that while degrees of belief are subject to some rational constraints (such as, but not limited to, the axioms of probability) these constraints usually do not determine a unique value. Rational people, in other words, may differ somewhat in their degrees of belief, even if they all have the same information.

PROPENSITY

Propensity theorists think of probability as a physical propensity, or disposition, or tendency of a given type of physical situation to yield an outcome of a certain kind or to yield a long run relative frequency of such an outcome. This kind of objective probability is sometimes called 'chance'.

Propensities, or chances, are not relative frequencies, but purported causes of the observed stable relative frequencies. Propensities are invoked to explain why repeating a certain kind of experiment will generate given outcome types at persistent rates, which are known as propensities or chances. Frequentists are unable to take this approach, since relative frequencies do not exist for single tosses of a coin, but only for large ensembles or collectives (see "single case possible" in the table above). In contrast, a propensitist is able to use the law of large numbers to explain the behaviour of long-run frequencies.

This law, which is a consequence of the axioms of probability, says that if (for example) a coin is tossed repeatedly many times, in such a way that its probability of landing heads is the same on each toss, and the outcomes are probabilistically independent, then the relative frequency of heads will be close to the probability of heads on each single toss. This law allows that stable long-run frequencies are a manifestation of invariant *single-case* probabilities. In addition to explaining the emergence of stable relative frequencies, the idea of propensity is motivated by the desire to make sense of single-case probability attributions in quantum mechanics, such as the probability of decay of a particular atom at a particular time.

The main challenge facing propensity theories is to say exactly what propensity means. (And then, of course, to show that propensity thus defined has the required properties.) At present, unfortunately, none of the well-recognised accounts of propensity comes close to meeting this challenge.

A propensity theory of probability was given by Charles Sanders Peirce. A later propensity theory was proposed by philosopher Karl Popper, who had only slight acquaintance with the writings of C. S. Peirce, however. Popper noted that the outcome of a physical experiment is produced by a certain set of "generating conditions". When we repeat an experiment, as the saying goes, we really perform another experiment with a (more or less) similar set of generating conditions. To say that a set of generating conditions has propensity p of producing the outcome E means that those exact conditions, if repeated indefinitely, would produce an outcome sequence in which E occurred with limiting relative frequency p. For Popper then, a deterministic experiment would have propensity 0 or 1 for each outcome, since those generating conditions would have same outcome on each trial. In other words, non-trivial propensities (those that differ from 0 and 1) only exist for genuinely indeterministic experiments.

A number of other philosophers, including David Miller and Donald A. Gillies, have proposed propensity theories somewhat similar to Popper`s.

Other propensity theorists (e.g. Ronald Giere) do not explicitly define propensities at all, but rather see propensity as defined by the theoretical role it plays in science. They argue, for example, that physical magnitudes such as electrical charge cannot be explicitly defined either, in terms of more basic things, but only in terms of what they do (such as attracting and repelling other electrical charges). In a similar way, propensity is whatever fills the various roles that physical probability plays in science.

What roles does physical probability play in science? What are its properties? One central property of chance is that, when known, it constrains rational belief to take the same numerical value. David Lewis called this the *Principal Principle*, (3.3 & 3.5) a term that philosophers have mostly adopted. For example, suppose you are certain that a particular biased coin has propensity 0.32 to land heads every time it is tossed. What is then the correct price for a gamble that pays $1 if the coin lands heads, and nothing otherwise? According to the Principal Principle, the fair price is 32 cents.

SUBJECTIVISM

Gambling odds reflect the average bettor's 'degree of belief' in the outcome.

Subjectivists, also known as **Bayesians** or followers of **epistemic probability**, give the notion of probability a subjective status by regarding it as a measure of the 'degree of belief' of the individual assessing the uncertainty of a particular situation. Epistemic or subjective probability is sometimes called **credence**, as opposed to the term **chance** for a propensity probability.

Some examples of epistemic probability are to assign a probability to the proposition that a proposed law of physics is true, and to determine how probable it is that a suspect committed a crime, based on the evidence presented.

Gambling odds don`t reflect the bookies` belief in a likely winner, so much as the other bettors` belief, because the bettors are actually betting against one another. The odds are set based on how many people have bet on a possible winner, so that even if the high odds players always win, the bookies will always make their percentages anyway.

The use of Bayesian probability raises the philosophical debate as to whether it can contribute valid justifications of belief.

Bayesians point to the work of Ramsey (p 182) and de Finetti (p 103) as proving that subjective beliefs must follow the laws of probability if they are to be coherent. Evidence casts doubt that humans will have coherent beliefs.

The use of Bayesian probability involves specifying a prior probability. This may be obtained from consideration of whether the required prior probability is greater or lesser than a reference probability associated with an urn model or a thought experiment. The issue is that for a given problem, multiple thought experiments could apply, and choosing one is a matter of judgement: different people may assign different prior probabilities, known as the reference class problem. The "sunrise problem" provides an example.

Prediction

An alternative account of probability emphasizes the role of *prediction* – predicting future observations on the basis of past observations, not on unobservable parameters. In its modern form, it is mainly in the Bayesian vein. This was the main function of probability before the 20th century, but fell out of favor compared to the parametric approach, which modeled phenomena as a physical system that was observed with error, such as in celestial mechanics. The modern predictive approach was pioneered by Bruno de Finetti, with the central idea of exchangeability – that future observations should behave like past observations. This view came to the attention of the Anglophone world with the 1974 translation of de Finetti`s book, and has since been propounded by such statisticians as Seymour Geisser.

BAYESIAN PROBABILITY

Bayesian probability is an interpretation of the concept of probability, in which, instead of frequency or propensity of some phenomenon, assigned probabilities represent states of knowledge or belief.

The Bayesian interpretation of probability can be seen as an extension of propositional logic that enables reasoning with hypotheses, i.e., the propositions whose truth or falsity is uncertain. In the Bayesian view, a probability is assigned to a hypothesis, whereas under frequentist inference, a hypothesis is typically tested without being assigned a probability.

Bayesian probability belongs to the category of evidential probabilities; to evaluate the probability of a hypothesis, the Bayesian probabilist specifies some prior probability, which is then updated to a posterior probability in the light of new, relevant data (evidence). The Bayesian interpretation provides a standard set of procedures and formulae to perform this calculation.

The term "Bayesian" derives from the 18th century mathematician and theologian Thomas Bayes, who provided the first mathematical treatment of a non-trivial problem of Bayesian inference. Mathematician Pierre-Simon Laplace pioneered and popularised what is now called Bayesian probability.

Broadly speaking, there are two views on Bayesian probability that interpret the *probability* concept in different ways. According to the *objectivist view*, the rules of Bayesian statistics can be justified by requirements of rationality and consistency and interpreted as an extension of logic. According to the *subjectivist view*, probability quantifies a "personal belief".

Bayesian Methodology

Bayesian methods are characterized by concepts and procedures as follows:

- The use of random variables or, more generally, unknown quantities, to model all sources of uncertainty in statistical models. This also includes uncertainty resulting from lack of information (see also aleatoric and epistemic uncertainty).
- The need to determine the *prior probability distribution* taking into account the available (prior) information.
- The *sequential use of Bayes' formula*: when more data become available, calculate the *posterior distribution* using Bayes' formula; subsequently, the posterior distribution becomes the next prior.
- While for the frequentist a hypothesis is a proposition (which must be either true or false), so that the frequentist probability of a hypothesis is either 0 or 1, in Bayesian statistics the probability that can be assigned to a hypothesis can also be in a range from 0 to 1 if the truth value is uncertain.

Objective and Subjective Bayesian Probabilities

Broadly speaking, there are two interpretations on Bayesian probability. For *objectivists, probability* objectively measures the plausibility of propositions, i.e., probability corresponds to a reasonable belief everyone (even a "robot") sharing the same knowledge should share in accordance with the rules of Bayesian statistics, which can be justified by requirements of rationality and consistency. For *subjectivists*, probability corresponds to a "personal belief"; rationality and consistency allow for substantial variation within the constraints they pose. The objective and subjective variants of Bayesian probability differ mainly in their interpretation and construction of the prior probability.

The term *Bayesian* refers to Thomas Bayes (1702–1761), who proved a special case of what is now called Bayes` theorem in a paper titled "An Essay towards solving a Problem in the Doctrine of Chances". In that special case, the prior and posterior distributions were Beta distributions and the data came from Bernoulli trials. It was Pierre-Simon Laplace (1749–1827) who introduced a general version of the theorem and used it to approach problems in celestial mechanics, medical statistics, reliability, and jurisprudence. Early Bayesian inference, which used uniform priors following Laplace`s principle of insufficient reason, was called "inverse probability" (because it infers backwards from observations to parameters, or from effects to causes). After the 1920s, "inverse probability" was largely supplanted by a collection of methods that came to be called frequentist statistics.

In the 20th century, the ideas of Laplace were further developed in two different directions, giving rise to *objective* and *subjective* currents in Bayesian practice. Harold Jeffreys` *Theory of Probability* (first published in 1939) played an important role in the revival of the Bayesian view of probability, followed by works by Abraham Wald (1950) and Leonard J. Savage (1954). The adjective *Bayesian* itself dates to the 1950s; the derived *Bayesianism, neo-Bayesianism* is of 1960s coinage. In the objectivist stream, the statistical analysis depends on only the model assumed and the data analysed. No subjective decisions need to be involved. In contrast, "subjectivist" statisticians deny the possibility of fully objective analysis for the general case.

In the 1980s there was a dramatic growth in research and applications of Bayesian methods, mostly attributed to the discovery of Markov chain Monte Carlo methods and the consequent removal of many of the computational problems, and to an increasing interest in nonstandard, complex applications. While frequentist statistics remains strong (as seen by the fact that most undergraduate teaching is still based on it), Bayesian methods are widely accepted and used, e.g., in the field of machine learning.

Justification of Bayesian Probabilities

The use of Bayesian probabilities as the basis of Bayesian inference has been supported by several arguments, such as Cox axioms, the Dutch book argument, arguments based on decision theory and de Finetti`s theorem.

Axiomatic Approach

Richard T. Cox showed that Bayesian updating follows from several axioms, including two functional equations and a hypothesis of differentiability. The assumption of differentiability or even continuity is controversial; Halpern found a counterexample based on his observation that the Boolean algebra of statements may be finite. Other axiomatizations have been suggested by various authors with the purpose of making the theory more rigorous.

Dutch book Approach

The Dutch book argument was proposed by de Finetti; it is based on betting. A Dutch book is made when a clever gambler places a set of bets that guarantee a profit, no matter what the outcome of the bets. If a bookmaker follows the rules of the Bayesian calculus in the construction of his odds, a Dutch book cannot be made.

However, Ian Hacking noted that traditional Dutch book arguments did not specify Bayesian updating: they left open the possibility that non-Bayesian updating rules could avoid Dutch books. For example, Hacking writes "And neither the Dutch book argument, nor any other in the personalist arsenal of proofs of the probability axioms, entails the dynamic assumption. Not one entails Bayesianism. So the personalist requires the dynamic assumption to be Bayesian. It is true that in consistency a personalist could abandon the Bayesian model of learning from experience. Salt could lose its savour."

In fact, there are non-Bayesian updating rules that also avoid Dutch books (as discussed in the literature on "probability kinematics" following the publication of Richard C. Jeffreys` rule, which is itself regarded as Bayesian). The additional hypotheses sufficient to (uniquely) specify Bayesian updating are substantial, complicated, and unsatisfactory.

Decision Theory Approach

A decision-theoretic justification of the use of Bayesian inference (and hence of Bayesian probabilities) was given by Abraham Wald, who proved that every admissible statistical procedure is either a Bayesian procedure or a limit of Bayesian procedures. Conversely, every Bayesian procedure is admissible.

Personal Probabilities and Objective Methods for Constructing Priors

Following the work on expected utility theory of Ramsey and von Neumann, decision-theorists have accounted for rational behavior using a probability distribution for the agent. Johann Pfanzagl completed the *Theory of Games and Economic Behavior* by providing an axiomatization of subjective probability and utility, a task left uncompleted by von Neumann and Oskar Morgenstern: their original theory supposed that all the agents had the same probability distribution, as a convenience. Pfanzagl`s axiomatization was endorsed by Oskar Morgenstern: "Von Neumann and I have anticipated" the question whether probabilities "might, perhaps more typically, be subjective and have stated specifically that in the latter case axioms could be found from which could derive the desired numerical utility together with a number for the probabilities (cf. p. 19 of The Theory of Games and Economic Behavior). We did not carry this out; it was demonstrated by Pfanzagl ... with all the necessary rigor".

Ramsey and Savage noted that the individual agent's probability distribution could be objectively studied in experiments. The role of judgment and disagreement in science has been recognized since Aristotle and even more clearly with Francis Bacon. The objectivity of science lies not in the psychology of individual scientists, but in the process of science and especially in statistical methods, as noted by C. S. Peirce. Recall that the objective methods for falsifying propositions about personal probabilities have been used for a half century, as noted previously. Procedures for testing hypotheses about probabilities (using finite samples) are due to Ramsey (1931) and de Finetti (1931, 1937, 1964, 1970). Both Bruno de Finetti and Frank P. Ramsey acknowledge their debts to pragmatic philosophy, particularly (for Ramsey) to Charles S. Peirce.

The "Ramsey test" for evaluating probability distributions is implementable in theory, and has kept experimental psychologists occupied for a half century. This work demonstrates that Bayesian-probability propositions can be falsified, and so meet an empirical criterion of Charles S. Peirce, whose work inspired Ramsey. (This falsifiability-criterion was popularized by Karl Popper.)

Modern work on the experimental evaluation of personal probabilities uses the randomization, blinding, and Boolean-decision procedures of the Peirce-Jastrow experiment. Since individuals act according to different probability judgments, these agents` probabilities are "personal" (but amenable to objective study).

Personal probabilities are problematic for science and for some applications where decision-makers lack the knowledge or time to specify an informed probability-distribution (on which they are prepared to act). To meet the needs of science and of human limitations, Bayesian statisticians have developed "objective" methods for specifying prior probabilities.

Indeed, some Bayesians have argued the prior state of knowledge defines *the* (unique) prior probability-distribution for "regular" statistical problems; cf. well-posed problems. Finding the right method for constructing such "objective" priors (for appropriate classes of regular problems) has been the quest of statistical theorists from Laplace to John Maynard Keynes, Harold Jeffreys, and Edwin Thompson Jaynes: These theorists and their successors have suggested several methods for constructing "objective" priors:

- Maximum entropy
- Transformation group analysis
- Reference analysis

Each of these methods contributes useful priors for "regular" one-parameter problems, and each prior can handle some challenging statistical models (with "irregularity" or several parameters). Each of these methods has been useful in Bayesian practice. Indeed, methods for constructing "objective" (alternatively, "default" or "ignorance") priors have been developed by avowed subjective (or "personal") Bayesians like James Berger (Duke University) and José-Miguel Bernardo (Universitat de València), simply because such priors are needed for Bayesian practice, particularly in science. The quest for "the universal method for constructing priors" continues to attract statistical theorists.

Thus, the Bayesian statistician needs either to use informed priors (using relevant expertise or previous data) or to choose among the competing methods for constructing "objective" priors.

FREQUENTIST PROBABILITY

John Venn.

Frequentist probability or **frequentism** is a standard interpretation of probability; it defines an event`s probability as the limit of its relative frequency in a large number of trials. This interpretation supports the statistical needs of experimental scientists and pollsters; probabilities can be found (in principle) by a repeatable objective process (and are thus ideally devoid of opinion). It does not support all needs; gamblers typically require estimates of the odds without experiments.

The development of the frequentist account was motivated by the problems and paradoxes of the previously dominant viewpoint, the classical interpretation. In the classical interpretation, probability was defined in terms of the principle of indifference, based on the natural symmetry of a problem, so, *e.g.* the probabilities of dice games arise from the natural symmetric 6-sidedness of the cube. This classical interpretation stumbled at any statistical problem that has no natural symmetry for reasoning.

Definition

In the frequentist interpretation, probabilities are discussed only when dealing with well-defined random experiments (or random samples). The set of all possible outcomes of a random experiment is called the sample space of the experiment. An event is defined as a particular subset of the sample space to be considered. For any given event, only one of two possibilities may hold: it occurs or it does not. The relative frequency of occurrence of an event, observed in a number of repetitions of the experiment, is a measure of the **probability** of that event. This is the core conception of probability in the frequentist interpretation.

Thus, if n_t is the total number of trials and n_x is the number of trials where the event x occurred, the probability $P(x)$ of the event occurring will be approximated by the relative frequency as follows:

$$P(x) \approx \frac{n_x}{n_t}.$$

Clearly, as the number of trials is increased, one might expect the relative frequency to become a better approximation of a "true frequency".

A claim of the frequentist approach is that in the "long run," as the number of trials approaches infinity, the relative frequency will converge *exactly* to the true probability:

$$P(x) = \lim_{n_t \to \infty} \frac{n_x}{n_t}.$$

Scope

The frequentist interpretation is a philosophical approach to the definition and use of probabilities; it is one of several such approaches. It does not claim to capture all connotations of the concept 'probable' in colloquial speech of natural languages.

As an interpretation, it is not in conflict with the mathematical axiomatization of probability theory; rather, it provides guidance for how to apply mathematical probability theory to real-world situations. It offers distinct guidance in the construction and design of practical experiments, especially when contrasted with the Bayesian interpretation. As to whether this guidance is useful, or is apt to mis-interpretation, has been a source of controversy. Particularly when the frequency interpretation of probability is mistakenly assumed to be the only possible basis for frequentist inference. So, for example, a list of mis-interpretations of the meaning of p-values accompanies the article on p-values; controversies are detailed in the article on statistical hypothesis testing. The Jeffreys–Lindley paradox shows how different interpretations, applied to the same data set, can lead to different conclusions about the 'statistical significance' of a result.

As William Feller noted:

There is no place in our system for speculations concerning the probability that the sun will rise tomorrow. Before speaking of it we should have to agree on an (idealized) model which would presumably run along the lines "out of infinitely many worlds one is selected at random..." Little imagination is required to construct such a model, but it appears both uninteresting and meaningless.

Feller`s comment was criticism of Laplace, who published a solution to the sunrise problem using an alternative probability interpretation. Despite Laplace`s explicit and immediate disclaimer in the source, based on expertise in astronomy as well as probability, two centuries of criticism have followed.

The frequentist view may have been foreshadowed by Aristotle, in *Rhetoric*, when he wrote: the probable is that which for the most part happens

Poisson clearly distinguished between objective and subjective probabilities in 1837. Soon thereafter a flurry of nearly simultaneous publications by Mill, Ellis ("On the Foundations of the Theory of Probabilities" and "Remarks on the Fundamental Principles of the Theory of Probabilities"), Cournot (*Exposition de la théorie des chances et des probabilités*)and Fries introduced the frequentist view. Venn provided a thorough exposition (*The Logic of Chance: An Essay on the Foundations and Province of the Theory of Probability* (published editions in 1866, 1876, 1888)) two decades later. These were further supported by the publications of Boole and Bertrand. By the end of the 19th century the frequentist interpretation was well established and perhaps dominant in the sciences. The following generation established the tools of classical inferential statistics (significance testing, hypothesis testing and confidence intervals) all based on frequentist probability.

Alternatively, Jacob Bernoulli (AKA James or Jacques) understood the concept of frequentist probability and published a critical proof (the weak law of large numbers) posthumously in *1713*. He is also credited with some appreciation for subjective probability (prior to and without Bayes theorem). Gauss and Laplace used frequentist (and other) probability in derivations of the least squares method a century later, a generation before Poisson. Laplace considered the probabilities of testimonies, tables of mortality, judgments of tribunals, etc. which are unlikely candidates for classical probability. In this view, Poisson`s contribution was his sharp criticism of the alternative "inverse" (subjective, Bayesian) probability interpretation. Any criticism by Gauss and Laplace was muted and implicit. (Their later derivations did not use inverse probability.)

Major contributors to "classical" statistics in the early 20th century included Fisher, Neyman and Pearson. Fisher contributed to most of statistics and made significance testing the core of experimental science; Neyman formulated confidence intervals and contributed heavily to sampling theory; Neyman and Pearson paired in the creation of hypothesis testing. All valued objectivity, so the best interpretation of probability available to them was frequentist. All were suspicious of "inverse probability" (the available alternative) with prior probabilities chosen by the using the principle of indifference. Fisher said, "... the theory of inverse probability is founded upon an error, [referring to Bayes theorem] and must be wholly rejected." (from his Statistical Methods for Research Workers). While Neyman was a pure frequentist, Fisher`s views of probability were unique; Both had nuanced view of probability. von Mises offered a combination of mathematical and philosophical support for frequentism in the era.

Etymology

According to the *Oxford English Dictionary*, the term 'frequentist' was first used by M. G. Kendall in 1949, to contrast with Bayesians, whom he called "non-frequentists". He observed

3....we may broadly distinguish two main attitudes. One takes probability as ‹a degree of rational belief›, or some similar idea...the second defines probability in terms of frequencies of occurrence of events, or by relative proportions in ‹populations› or ‹collectives›; (p. 101)

...

12. It might be thought that the differences between the frequentists and the non-frequentists (if I may call them such) are largely due to the differences of the domains which they purport to cover. (p. 104)

...

I assert that this is not so ... The essential distinction between the frequentists and the non-frequentists is, I think, that the former, in an effort to avoid anything savouring of matters of opinion, seek to define probability in terms of the objective properties of a population, real or hypothetical, whereas the latter do not. [emphasis in original]

"The Frequency Theory of Probability" was used a generation earlier as a chapter title in Keynes (1921).

The historical sequence: probability concepts were introduced and much of probability mathematics derived (prior to the 20th century), classical statistical inference methods were developed, the mathematical foundations of probability were solidified and current terminology was introduced (all in the 20th century). The primary historical sources in probability and statistics did not use the current terminology of classical, subjective (Bayesian) and frequentist probability.

Alternative Views

Probability theory is a branch of mathematics. While its roots reach centuries into the past, it reached maturity with the axioms of Andrey Kolmogorov in 1933. The theory focuses on the valid operations on probability values rather than on the initial assignment of values; the mathematics is largely independent of any interpretation of probability.

Applications and interpretations of probability are considered by philosophy, the sciences and statistics. All are interested in the extraction of knowledge from observations—inductive reasoning. There are a variety of competing interpretations; All have problems. Major interpretations include classical probability, subjective probability and frequency interpretations.

- Classical probability assigns probabilities based on physical idealized symmetry (dice, coins, cards). The classical definition is at risk of circularity; Probabilities are defined by assuming equality of probabilities. In the absence of symmetry the utility of the definition is limited.
- Subjective (Bayesian) probability (a family of competing interpretations) considers degrees of belief. All practical "subjective" probability interpretations are so

constrained to rationality as to avoid most subjectivity. Real subjectivity is repellent to the sciences which strive for results independent of the observer and analyst. The historical roots of this concept extended to such non-numeric applications as legal evidence.

- Frequency interpretations are empirical—they are defined by a ratio from an infinite series of trials. This is a very natural interpretation for scientific experiments. Mathematicians are dubious of the convergence properties of the non-mathematical series.

The frequentist interpretation does resolve difficulties with the classical interpretation, such as any problem where the natural symmetry of outcomes is not known. It does not address other issues, such as the dutch book. Propensity probability is an alternative physicalist approach.

REFERENCES

1. Neyman, Jerzy (30 August 1937). "Outline of a Theory of Statistical Estimation Based on the Classical Theory of Probability". Phil. Trans. R. Soc. Lond. A. **236**: 333–380. doi:10.1098/rsta.1937.0005. Neyman`s derivation of confidence intervals embraced the measure theoretic axioms of probability published by Kolmogorov a few years previously and referenced the subjective (Bayesian) probability definitions of Jeffreys published earlier in the decade. Neyman defined frequentist probability (under the name classical) and stated the need for randomness in the repeated samples or trials. He accepted in principle the possibility of multiple competing theories of probability while expressing several specific reservations about the existing alternative probability interpretation.
2. von Mises, Richard (1939) Probability, Statistics, and Truth (in German) (English translation, 1981: Dover Publications; 2 Revised edition. ISBN 0486242145) (p.14)
3. William Feller (1957), An Introduction to Probability Theory and Its Applications, Vol. 1, page 4
4. Keynes, John Maynard; A Treatise on Probability (1921), Chapter VIII "The Frequency Theory of Probability".
5. Rhetoric Bk 1 Ch 2; discussed in J. Franklin, The Science of Conjecture: Evidence and Probability Before Pascal (2001), The Johns Hopkins University Press. ISBN 0801865697 , p. 110.
6. Gigerenzer, Gerd; Swijtink, Porter; Daston, Beatty; Daston, Krüger (1989). The Empire of chance : how probability changed science and everyday life. Cambridge Cambridgeshire New York: Cambridge University Press. pp. 35–6, 45. ISBN 0-521-39838-X.
7. Ellis, Robert Leslie (1843) "On the Foundations of the Theory of Probabilities", Transactions of the Cambridge Philosophical Society vol 8

8. Ellis, Robert Leslie (1854) "Remarks on the Fundamental Principles of the Theory of Probabilitiess", Transactions of the Cambridge Philosophical Society vol 9

9. Cournot, Antoine Augustin (1843) Exposition de la théorie des chances et des probabilités. L. Hachette, Paris. archive.org

10. Venn, John (1888) The Logic of Chance, 3rd Edition archive.org. Full title: The Logic of Chance: An essay on the foundations and province of the theory of probability, with especial reference to its logical bearings and its application to Moral and Social Science, and to Statistics, Macmillan & Co, London

11. Hald, Anders (2004). A history of parametric statistical inference from Bernoulli to Fisher, 1713 to 1935. København: Anders Hald, Department of applied Mathematics and Statistics, University of Copenhagen. pp. 11–12. ISBN 87-7834-628-2.

12. Fienberg, Stephen E. (1992). "A Brief History of Statistics in Three and One-half Chapters: A Review Essay". Statistical Science. 7 *(2): 208–225. doi:10.1214/* ss/1177011360.

13. David, F. N. (1962). Games, Gods & Gambling. New York: Hafner. pp. 137–138. Bernoulli provided a classical example of drawing a large number of black and white pebbles from an urn (with replacement). The sample ratio allowed Bernoulli to infer the ratio in the urn, with tighter bounds as the number of samples increased. Historians can interpret the example as classical, frequentist or subjective probability. David says, "James has definitely started here the controversy on inverse probability..." Bernoulli wrote generations before Bayes, LaPlace and Gauss. The controversy continues.

14. Hald, Anders (2004). A history of parametric statistical inference from Bernoulli to Fisher, 1713 to 1935. København: Anders Hald, Department of applied Mathematics and Statistics, University of Copenhagen. pp. 1–5. ISBN 87-7834-628-2.

15. The Frequency theory Chapter 5; discussed in Donald Gilles, Philosophical theories of probability (2000), Psychology Press. ISBN 9780415182751 , p. 88.

16. Earliest Known Uses of Some of the Words of Probability & Statistics

17. Kendall, Maurice George (1949). "On the Reconciliation of Theories of Probability". Biometrika. Biometrika Trust. **36** *(1/2): 101–116. doi:10.1093/* biomet/36.1-2.101. JSTOR 2332534.

18. Hájek, Alan, Zalta, Edward N., ed., Interpretations of Probability, The Stanford Encyclopedia of Philosophy

19. Ash, Robert B. (1970). Basic Probability Theory. New York: Wiley. pp. 1–2.

CHAPTER 3

Probability Distributions

Chapter Contents

INTRODUCTION

Probability theory is the branch of mathematics concerned with probability, the analysis of random phenomena. The central objects of probability theory are random variables, stochastic processes, and events: mathematical abstractions of non-deterministic events or measured quantities that may either be single occurrences or evolve over time in an apparently random fashion.

It is not possible to predict precisely results of random events. However, if a sequence of individual events, such as coin flipping or the roll of dice, is influenced by other factors, such as friction, it will exhibit certain patterns, which can be studied and predicted. Two representative mathematical results describing such patterns are the law of large numbers and the central limit theorem.

As a mathematical foundation for statistics, probability theory is essential to many human activities that involve quantitative analysis of large sets of data.Methods of probability theory also apply to descriptions of complex systems given only partial knowledge of their state, as in statistical mechanics. A great discovery of twentieth century physics was the probabilistic nature of physical phenomena at atomic scales, described in quantum mechanics.

HISTORY

The mathematical theory of probability has its roots in attempts to analyze games of chance by Gerolamo Cardano in the sixteenth century, and by Pierre de Fermat and Blaise Pascal in the seventeenth century (for example the "problem of points). Christiaan Huygens published a book on the subject in 1657 and in the 19th century, Pierre Laplace completed what is today considered the classic interpretation.

Initially, probability theory mainly considered **discrete** events, and its methods were mainly combinatorial. Eventually, analytical considerations compelled the incorporation of **continuous** variables into the theory.

This culminated in modern probability theory, on foundations laid by Andrey Nikolaevich Kolmogorov. Kolmogorov combined the notion of sample space, introduced by Richard von Mises, and **measure theory** and presented his axiom system for probability theory in 1933. This became the mostly undisputed axiomatic basis for modern probability theory; but, alternatives exist, such as the adoption of finite rather than countable additivity by Bruno de Finetti.

TREATMENT

Most introductions to probability theory treat discrete probability distributions and continuous probability distributions separately. The more mathematically advanced measure theory-based treatment of probability covers the discrete, continuous, a mix of the two, and more.

Motivation

Consider an experiment that can produce a number of outcomes. The set of all outcomes is called the sample space of the experiment. The power set of the sample space (or equivalently, the event space) is formed by considering all different collections of possible results. For example, rolling an honest die produces one of six possible results. One collection of possible results corresponds to getting an odd number. Thus, the subset {1,3,5} is an element of the power set of the sample space of die rolls. These collections are called events. In this case, {1,3,5} is the event that the die falls on some odd number. If the results that actually occur fall in a given event, that event is said to have occurred.

Probability is a way of assigning every "event" a value between zero and one, with the requirement that the event made up of all possible results (in our example, the event {1,2,3,4,5,6}) be assigned a value of one. To qualify as a probability distribution, the assignment of values must satisfy the requirement that if you look at a collection of mutually exclusive events (events that contain no common results, e.g., the events {1,6}, {3}, and {2,4} are all mutually exclusive), the probability that one of the events will occur is given by the sum of the probabilities of the individual events.

The probability that any one of the events {1,6}, {3}, or {2,4} will occur is 5/6. This is the same as saying that the probability of event {1,2,3,4,6} is 5/6. This event encompasses the possibility of any number except five being rolled. The mutually exclusive event {5} has a probability of 1/6, and the event {1,2,3,4,5,6} has a probability of 1, that is, absolute certainty.

Discrete Probability Distributions

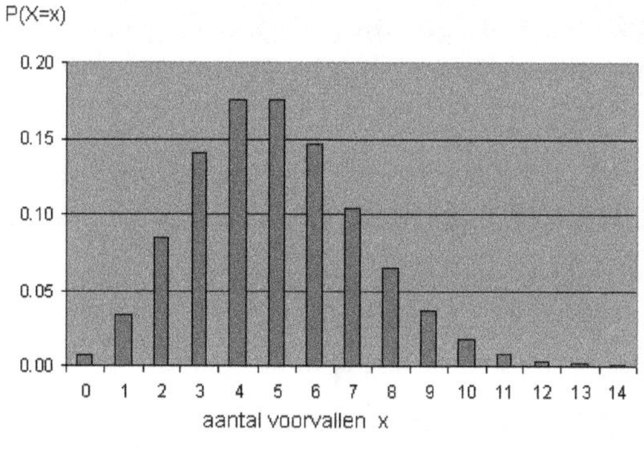

The Poisson distribution, a discrete probability distribution.

Discrete Probability Theory deals with events that occur in countable sample spaces.

Examples: Throwing dice, experiments with decks of cards, random walk, and tossing coins

Classical definition: Initially the probability of an event to occur was defined as number of cases favorable for the event, over the number of total outcomes possible in an equiprobable sample space: see Classical definition of probability.

For example, if the event is occurrence of an even number when a die is rolled, the probability is given by $\frac{3}{6}=\frac{1}{2}$, since 3 faces out of the 6 have even numbers and each face has the same probability of appearing.

Modern definition: The modern definition starts with a finite or countable set called the **sample space**, which relates to the set of all possible outcomes in classical sense, denoted by Ω. It is then assumed that for each element $x \in \Omega$, an intrinsic "probability" value $f(x)$ is attached, which satisfies the following properties:

1. $f(x) \in [0,1]$ for all $x \in \Omega$;
2. $\sum_{x\in\Omega} f(x) = 1$.

That is, the probability function f(x) lies between zero and one for every value of x in the sample space Ω, and the sum of f(x) over all values x in the sample space Ω is equal to 1. An **event** is defined as any subset E of the sample space Ω. The **probability** of the event E is defined as

$$P(E) = \sum_{x \in E} f(x).$$

So, the probability of the entire sample space is 1, and the probability of the null event is 0.

The function $f(x)$ mapping a point in the sample space to the "probability" value is called a **probability mass function** abbreviated as **pmf**. The modern definition does not try to answer how probability mass functions are obtained; instead it builds a theory that assumes their existence.

Continuous Probability Distributions

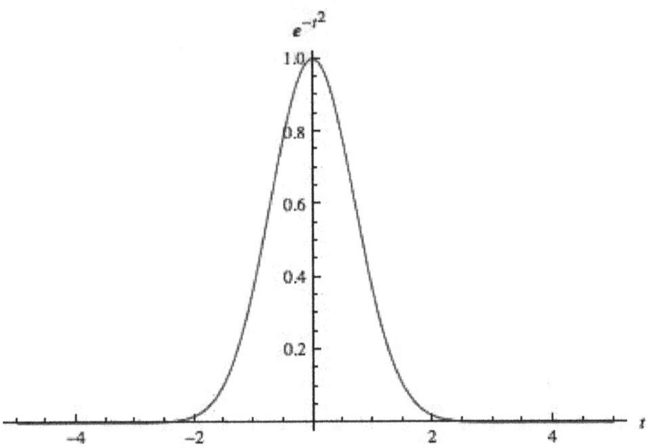

The normal distribution, a continuous probability distribution.

Continuous probability theory deals with events that occur in a continuous sample space.

Classical definition: The classical definition breaks down when confronted with the continuous case. See Bertrand›s paradox.

Modern definition: If the outcome space of a random variable X is the set of real numbers $(\mathbb{R})$ or a subset thereof, then a function called the **cumulative distribution function** (or **cdf**) F exists, defined by $F(x) = P(X \le x)$. That is, F(x) returns the probability that X will be less than or equal to x.

The cdf necessarily satisfies the following properties.

- F is a monotonically non-decreasing, right-continuous function;
- $\lim_{x \to -\infty} F(x) = 0$;

$$\lim_{x\to\infty} F(x) = 1 .$$

■

If F is absolutely continuous, i.e., its derivative exists and integrating the derivative gives us the cdf back again, then the random variable X is said to have a **probability density function** or **pdf** or simply **density** $f(x) = \frac{dF(x)}{dx}$

For a set $E \subseteq \mathbb{R}$, the probability of the random variable X being in E is

$$P(X \in E) = \int_{x\in E} dF(x) .$$

In case the probability density function exists, this can be written as

$$P(X \in E) = \int_{x\in E} f(x)\, dx .$$

Whereas the pdf exists only for continuous random variables, the cdf exists for all random variables (including discrete random variables) that take values in $\mathbb{R}$

These concepts can be generalized for multidimensional cases on $\mathbb{R}^n$ and other continuous sample spaces.

Measure-Theoretic Probability Theory

The raison d'être of the measure-theoretic treatment of probability is that it unifies the discrete and the continuous cases, and makes the difference a question of which measure is used. Furthermore, it covers distributions that are neither discrete nor continuous nor mixtures of the two.

An example of such distributions could be a mix of discrete and continuous distributions—for example, a random variable that is 0 with probability 1/2, and takes a random value from a normal distribution with probability 1/2. It can still be studied to some extent by considering it to have a pdf of is the Dirac delta function.

Other distributions may not even be a mix, for example, the Cantor distribution has no positive probability for any single point, neither does it have a density. The modern approach to probability theory solves these problems using measure theory to define the probability space:

Given any set Ω, (also called **sample space**) and a σ-algebra F on it, a measure P on F is called a **probability measure** if $P(\Omega) = 1$.

If F is the Borel σ-algebra on the set of real numbers, then there is a unique probability measure on $\mathcal{F}$ for any cdf, and vice versa. The measure corresponding to a cdf is said to be **induced** by the cdf. This measure coincides with the pmf for discrete variables and pdf for continuous variables, making the measure-theoretic approach free of fallacies.

The probability of a set E in the σ-algebra $\mathcal{F}$ is defined as

$$P(E) = \int_{\omega \in E} \mu_F(d\omega)$$

where the integration is with respect to the measure μ_F induced by F

Along with providing better understanding and unification of discrete and continuous probabilities, measure-theoretic treatment also allows us to work on probabilities outside $\mathbb{R}^n$, as in the theory of stochastic processes. For example, to study Brownian motion, probability is defined on a space of functions.

When it's convenient to work with a dominating measure, the Radon-Nikodym theorem is used to define a density as the Radon-Nikodym derivative of the probability distribution of interest with respect to this dominating measure. Discrete densities are usually defined as this derivative with respect to a counting measure over the set of all possible outcomes. Densities for absolutely continuous distributions are usually defined as this derivative with respect to the Lebesgue measure. If a theorem can be proved in this general setting, it holds for both discrete and continuous distributions as well as others; separate proofs are not required for discrete and continuous distributions.

CLASSICAL PROBABILITY DISTRIBUTIONS

Certain random variables occur very often in probability theory because they well describe many natural or physical processes. Their distributions therefore have gained special importance in probability theory. Some fundamental discrete distributions are the discrete uniform, Bernoulli, binomial, negative binomial, Poisson and geometric distributions. Important continuous distributions include the continuous uniform, normal, exponential, gamma and beta distributions.

CONVERGENCE OF RANDOM VARIABLES

In probability theory, there are several notions of convergence for random variables. They are listed below in the order of strength, i.e., any subsequent notion of convergence in the list implies convergence according to all of the preceding notions.

Weak Convergence

A sequence of random variables $X_1, X_2, \ldots,$ converges **weakly** to the random variable X if their respective cumulative distribution functions $F_1, F_2, \ldots$ converge to the cumulative distribution function F of X, wherever F is continuous. Weak convergence is also called **convergence in distribution**.

Most common shorthand notation: $X_n \xrightarrow{\mathcal{D}} X$

Convergence in Probability

The sequence of random variables $X_1, X_2, \dots$ is said to converge towards the random variable X **in probability** if $\lim_{n\to\infty} P(|X_n - X| \geq \varepsilon) = 0$ for every ε > 0.

Most common shorthand notation: $X_n \xrightarrow{P} X$

Strong convergence

The sequence of random variables $X_1, X_2, \dots$ is said to converge towards the random variable X **strongly** if $P(\lim_{n\to\infty} X_n = X) = 1$. Strong convergence is also known as **almost sure convergence**.

Most common shorthand notation: $X_n \xrightarrow{\text{a.s.}} X$

As the names indicate, weak convergence is weaker than strong convergence. In fact, strong convergence implies convergence in probability, and convergence in probability implies weak convergence. The reverse statements are not always true.

Law of Large Numbers

Common intuition suggests that if a fair coin is tossed many times, then roughly half of the time it will turn up heads, and the other half it will turn up tails. Furthermore, the more often the coin is tossed, the more likely it should be that the ratio of the number of heads to the number of tails will approach unity. Modern probability theory provides a formal version of this intuitive idea, known as the **law of large numbers**. This law is remarkable because it is not assumed in the foundations of probability theory, but instead emerges from these foundations as a theorem. Since it links theoretically derived probabilities to their actual frequency of occurrence in the real world, the law of large numbers is considered as a pillar in the history of statistical theory and has had widespread influence.

The **law of large numbers** (LLN) states that the sample average

$$\overline{X}_n = \frac{1}{n}\sum_{k=1}^{n} X_k$$

of a sequence of independent and identically distributed random variables X_k converges towards their common expectation 'μ', provided that the expectation of $|X_k|$ is finite.

It is in the different forms of convergence of random variables that separates the weak and the strong law of large numbers

Weak law: $\overline{X}_n \xrightarrow{P} \mu$ for $n \to \infty$

Strong law: $\overline{X}_n \xrightarrow{\text{a. s.}} \mu$ for $n \to \infty$.

It follows from the LLN that if an event of probability p is observed repeatedly during independent experiments, the ratio of the observed frequency of that event to the total number of repetitions converges towards p.

For example, if $Y_1, Y_2, \ldots$ are independent Bernoulli random variables taking values 1 with probability p and 0 with probability 1-p, then $\mathrm{E}(Y_i) = p$ for all i, so that $\bar{Y}_n$ converges to p almost surely.

Central Limit Theorem

"The central limit theorem (CLT) is one of the great results of mathematics." (Chapter 18 in) It explains the ubiquitous occurrence of the normal distribution in nature.

The theorem states that the average of many independent and identically distributed random variables with finite variance tends towards a normal distribution irrespective of the distribution followed by the original random variables. Formally, let $X_1, X_2, \ldots$ be independent random variables with mean μ and variance $\sigma^2 > 0$ Then the sequence of random variables

$$Z_n = \frac{\sum_{i=1}^n (X_i - \mu)}{\sigma\sqrt{n}}$$

converges in distribution to a standard normal random variable.

For some classes of random variables the classic central limit theorem works rather fast (see Berry–Esseen theorem), for example the distributions with finite first, second and third moment from the exponential family; on the other hand, for some random variables of the heavy tail and fat tail variety, it works very slowly or may not work at all: in such cases one may use the Generalized Central Limit Theorem (GCLT).

PROBABILITY DISTRIBUTION

In probability and statistics, a **probability distribution** is a mathematical function that, stated in simple terms, can be thought of as providing the probability of occurrence of different possible outcomes in an experiment. For instance, if the random variable X is used to denote the outcome of a coin toss ('the experiment'), then the probability distribution of X would take the value 0.5 for $X = \text{Heads}$, and 0.5 for $X = \text{Tails}$.

In more technical terms, the probability distribution is a description of a random phenomenon in terms of the probabilities of events. Examples of random phenomena can include the results of an experiment or survey. A probability distribution is defined in terms of an underlying sample space, which is the set of all possible outcomes of the random phenomenon being observed. The sample space may be the set of real numbers or a higher-dimensional vector space, or it may be a list of non-numerical values; for example, the sample space of a coin flip would be $\{\text{Heads}, \text{Tails}\}$.

Probability distributions are generally divided into two classes. A **discrete probability distribution** (applicable to the scenario where the set of possible outcomes is discrete, such as a coin toss or a roll of dice) can be encoded by a discrete list of the probabilities of the outcomes, known as a probability mass function. On the other hand, a **continuous probability distribution** (applicable to the scenarios where the set of possible outcomes can take on values in a continuous range (e.g., real numbers), such as the temperature on a given day) is typically described by probability density functions (with the probability of any individual outcome actually being 0). The normal distribution represents a commonly encountered continuous probability distribution. More complex experiments, such as those involving stochastic processes defined in continuous time, may demand the use of more general probability measures.

A probability distribution whose sample space is the set of real numbers is called univariate, while a distribution whose sample space is a vector space is called multivariate. A univariate distribution gives the probabilities of a single random variable taking on various alternative values; a multivariate distribution (a joint probability distribution) gives the probabilities of a random vector—a list of two or more random variables—taking on various combinations of values. Important and commonly encountered univariate probability distributions include the binomial distribution, the hypergeometric distribution, and the normal distribution. The multivariate normal distribution is a commonly encountered multivariate distribution.

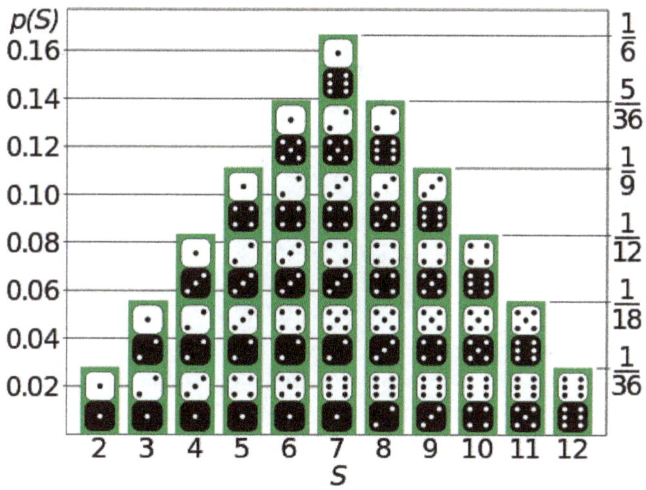

The probability mass function (pmf) p(S) specifies the probability distribution for the sum S of counts from two dice. For example, the figure shows that p(11) = 1/18. The pmf allows the computation of probabilities of events such as P(S > 9) = 1/12 + 1/18 + 1/36 = 1/6, and all other probabilities in the distribution.

To define probability distributions for the simplest cases, one needs to distinguish between **discrete** and **continuous** random variables. In the discrete case, it is sufficient to specify a probability mass function assigning a probability to each possible outcome: for example, when throwing a fair dice, each of the six values 1 to 6 has the probability 1/6. The probability of an event is then defined to be the sum of the probabilities of the outcomes that satisfy the event; for example, the probability of the event "the die rolls an even value" is

$$\mathbf{Prob(2) + Prob(4) + Prob(6) = 1/6 + 1/6 + 1/6 = 1/2.}$$

In contrast, when a random variable takes values from a continuum then typically, any individual outcome has probability zero and only events that include infinitely many outcomes, such as intervals, can have positive probability. For example, the probability that a given object weighs exactly 500 g is zero, because the probability of measuring exactly 500 g tends to zero as the accuracy of our measuring instruments increases. Nevertheless, in quality control one might demand that the probability of a "500 g" package containing between 490 g and 510 g should be no less than 98%, and this demand is less sensitive to the accuracy of our instruments.

Continuous probability distributions can be described in several ways. The probability density function describes the infinitesimal probability of any given value, and the probability that the outcome lies in a given interval can be computed by integrating the probability density function over that interval. On the other hand, the cumulative distribution function describes the probability that the random variable is no larger than a given value; the probability that the outcome lies in a given interval can be computed by taking the difference between the values of the cumulative distribution function at the endpoints of the interval. The cumulative distribution function is the antiderivative of the probability density function provided that the latter function exists.

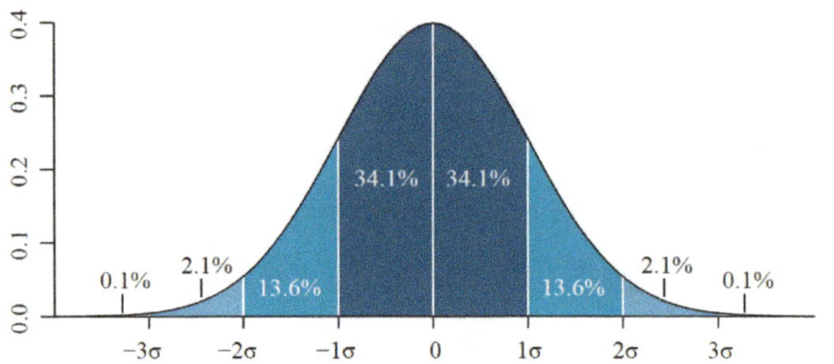

The probability density function (pdf) of the normal distribution, also called Gaussian or bell curve, the most important continuous random distribution. As notated on the figure, the probabilities of intervals of values correspond to the area under the curve.

Terminology

As probability theory is used in quite diverse applications, terminology is not uniform and sometimes confusing. The following terms are used for non-cumulative probability distribution functions:

- **Probability mass**, Probability mass function, **p.m.f.**: for discrete random variables.
- Categorical distribution: for discrete random variables with a finite set of values.

- **Probability density**, Probability density function, **p.d.f.**: most often reserved for continuous random variables.

The following terms are somewhat ambiguous as they can refer to non-cumulative or cumulative distributions, depending on authors' preferences:

- **Probability distribution function**: continuous or discrete, non-cumulative or cumulative.
- **Probability function**: even more ambiguous, can mean any of the above or other things.

Finally,

- **Probability distribution**: sometimes the same as probability distribution function, but usually refers to the more complete assignment of probabilities to all measurable subsets of outcomes (i.e. the corresponding probability measure), not just to specific outcomes or ranges of outcomes.

Basic Terms

- Mode: for a discrete random variable, the value with highest probability (the location at which the probability mass function has its peak); for a continuous random variable, the location at which the probability density function has its peak.
- Support: the smallest closed set whose complement has probability zero.
- Head: the range of values where the pmf or pdf is relatively high.
- Tail: the complement of the head within the support; the large set of values where the pmf or pdf is relatively low.
- Expected value or **mean**: the weighted average of the possible values, using their probabilities as their weights; or the continuous analog thereof.
- Median: the value such that the set of values less than the median has a probability of one-half.
- Variance: the second moment of the pmf or pdf about the mean; an important measure of the dispersion of the distribution.
- Standard deviation: the square root of the variance, and hence another measure of dispersion.
- **Symmetry**: a property of some distributions in which the portion of the distribution to the left of a specific value is a mirror image of the portion to its right.
- Skewness: a measure of the extent to which a pmf or pdf "leans" to one side of its mean. The third standardized moment of the distribution.
- Kurtosis: a measure of the fatness of the tails of a pmf or pdf. The fourth standardized moment of the distribution.

Cumulative Distribution Function

Because a probability distribution Pr on the real line is determined by the probability of a scalar random variable X being in a half-open interval (-∞, x], the probability distribution is completely characterized by its cumulative distribution function:

$F(x) = \Pr[X \leq x] \qquad \text{for all } x \in \mathbb{R}.$

Discrete Probability Distribution

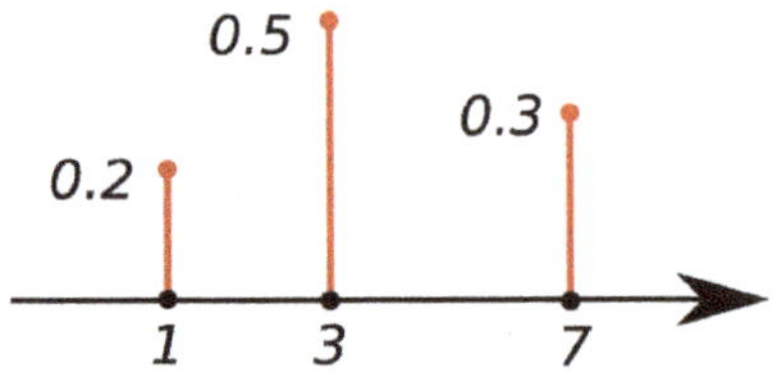

The probability mass function of a discrete probability distribution. The probabilities of the singletons {1}, {3}, and {7} are respectively 0.2, 0.5, 0.3. A set not containing any of these points has probability zero.

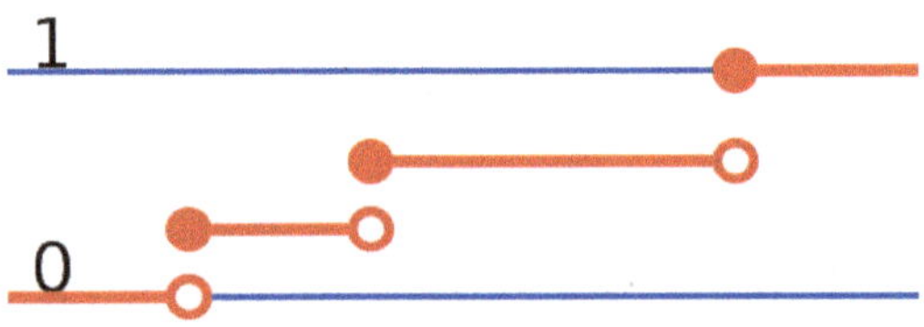

The cdf of a discrete probability distribution, ...

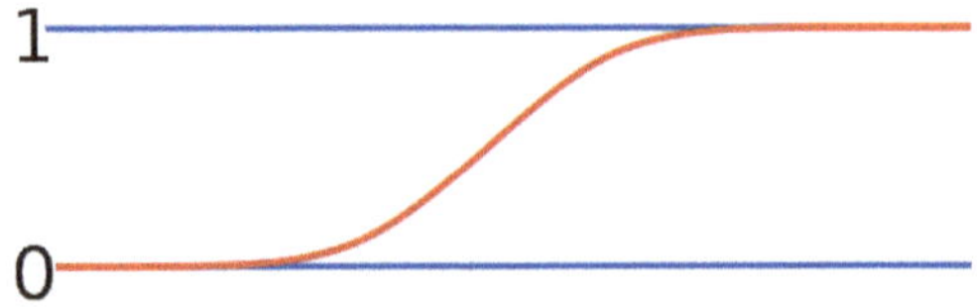

... of a continuous probability distribution, ...

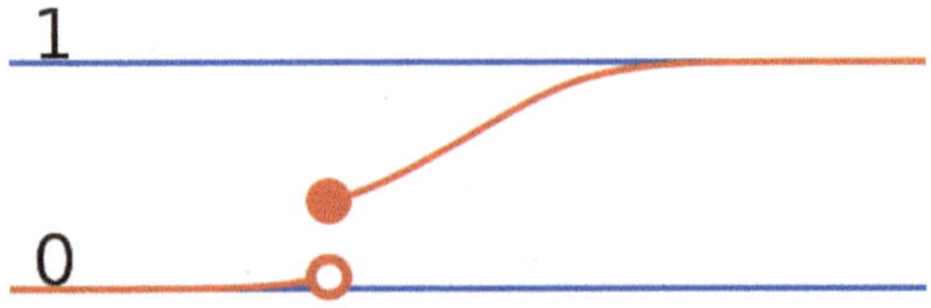

... of a distribution which has both a continuous part and a discrete part.

A **discrete probability distribution** is a probability distribution characterized by a probability mass function. Thus, the distribution of a random variable X is discrete, and X is called a **discrete random variable**, if

$$\sum_u \Pr(X = u) = 1$$

as u runs through the set of all possible values of X. A discrete random variable can assume only a finite or countably infinite number of values. For the number of potential values to be countably infinite, even though their probabilities sum to 1, the probabilities have to decline to zero fast enough. For example, if $\Pr(X = n) = \frac{1}{2^n}$ for $n = 1, 2, \ldots,$ we have the sum of probabilities 1/2 + 1/4 + 1/8 + ... = 1.

Well-known discrete probability distributions used in statistical modeling include the Poisson distribution, the Bernoulli distribution, the binomial distribution, the geometric distribution, and the negative binomial distribution. Additionally, the discrete uniform distribution is commonly used in computer programs that make equal-probability random selections between a number of choices.

Measure Theoretic Formulation

A measurable function $X: A \to B$ between a probability space $(A, \mathcal{A}, P)$ and a measurable space $(B, \mathcal{B})$ is called a **discrete random variable** provided its image is a countable set and the pre-image of singleton sets are measurable, i.e., $X^{-1}(b) \in \mathcal{A}$ for all $b \in B$. The latter requirement induces a probability mass function $f_X: X(A) \to \mathbb{R}$ via $f_X(b) := P(X^{-1}(b))$. Since the pre-images of disjoint sets are disjoint

$$\sum_{b \in X(A)} f_X(b) = \sum_{b \in X(A)} P(X^{-1}(b)) = P\left(\bigcup_{b \in X(A)} X^{-1}(b)\right) = P(A) = 1.$$

This recovers the definition given above.

Cumulative Density

Equivalently to the above, a discrete random variable can be defined as a random variable whose cumulative distribution function (cdf) increases only by jump discontinuities—that is, its cdf increases only where it jumps to a higher value, and is constant between those jumps. The points where jumps occur are precisely the values which the random variable may take.

Delta-function Representation

Consequently, a discrete probability distribution is often represented as a generalized probability density function involving Dirac delta functions, which substantially unifies the treatment of continuous and discrete distributions. This is especially useful when dealing with probability distributions involving both a continuous and a discrete part.

Indicator-function Representation

For a discrete random variable X, let u_0, u_1, ... be the values it can take with non-zero probability. Denote

$\Omega_i = X^{-1}(u_i) = \{\omega : X(\omega) = u_i\},\ i = 0, 1, 2, \ldots$

These are disjoint sets, and by formula (1)

$$\Omega_i = X^{-1}(u_i) = \{\omega : X(\omega) = u_i\},\ i = 0, 1, 2, \ldots$$

It follows that the probability that X takes any value except for u_0, u_1, ... is zero, and thus one can write X as

$$X(\omega) = \sum_i u_i 1_{\Omega_i}(\omega)$$

except on a set of probability zero, where A is the indicator function of A. This may serve as an alternative definition of discrete random variables.

Continuous Probability Distribution

A **continuous probability distribution** is a probability distribution that has a cumulative distribution function that is continuous. Most often they are generated by having a probability density function. Mathematicians call distributions with probability density functions **absolutely continuous**, since their cumulative distribution function is absolutely continuous with respect to the Lebesgue measure λ. If the distribution of X is continuous, then X is called a **continuous random variable**. There are many examples of continuous probability distributions: normal, uniform, chi-squared, and others.

Intuitively, a continuous random variable is the one which can take a continuous range of values—as opposed to a discrete distribution, where the set of possible values for the random variable is at most countable. While for a discrete distribution an event with probability zero is impossible (e.g., rolling 31/2 on a standard dice is impossible, and has probability zero), this is not so in the case of a continuous random variable. For example, if one measures the width of an oak leaf, the result of 3½ cm is possible; however, it has probability zero because uncountably many other potential values exist even between 3 cm and 4 cm. Each of these individual outcomes has probability zero, yet the probability that the outcome will fall into the interval (3 cm, 4 cm) is nonzero. This apparent paradox is resolved by the fact that the probability that X attains some value within an infinite set, such as an interval, cannot be found by naively adding the probabilities for individual values. Formally, each value has an infinitesimally small probability, which statistically is equivalent to zero.

Formally, if X is a continuous random variable, then it has a probability density function $f(x)$, and therefore its probability of falling into a given interval, say [a, b] is given by the integral

$$\Pr[a \le X \le b] = \int_a^b f(x)\,dx$$

In particular, the probability for X to take any single value a (that is $a \le X \le a$) is zero, because an integral with coinciding upper and lower limits is always equal to zero.

The definition states that a continuous probability distribution must possess a density, or equivalently, its cumulative distribution function be absolutely continuous. This requirement is stronger than simple continuity of the cumulative distribution function, and there is a special class of distributions, singular distributions, which are neither continuous nor discrete nor a mixture of those. An example is given by the Cantor distribution. Such singular distributions however are never encountered in practice.

Note on terminology: some authors use the term continuous distribution to denote the distribution with continuous cumulative distribution function. Thus, their definition includes both the (absolutely) continuous and singular distributions.

By one convention, a probability distribution μ is called continuous if its cumulative distribution function $F(x) = \mu(-\infty, x]$ is continuous and, therefore, the probability measure of singletons $\mu\{x\} = 0$ for all x.

Another convention reserves the term continuous probability distribution for absolutely continuous distributions. These distributions can be characterized by a probability density function: a non-negative Lebesgue integrable function {\displaystyle \,f} defined on the real numbers such that

$$F(x) = \mu(-\infty, x] = \int_{-\infty}^{x} f(t)\,dt.$$

Discrete distributions and some continuous distributions (like the Cantor distribution) do not admit such a density.

Some Properties

- The probability distribution of the sum of two independent random variables is the **convolution** of each of their distributions.
- Probability distributions are not a vector space—they are not closed under linear combinations, as these do not preserve non-negativity or total integral 1—but they are closed under convex combination, thus forming a convex subset of the space of functions (or measures).

Kolmogorov Definition

In the measure-theoretic formalization of probability theory, a random variable is defined as a measurable function X from a probability space $(\Omega,\mathcal{F},P)$ to measurable space $(\mathcal{X},\mathcal{A})$. A **probability distribution of *X*** is the pushforward measure X_*P of X , which is a probability measure on $X_*P = PX^{-1}$ satisfying $X_*P = PX^{-1}$.

Random Number Generation

A frequent problem in statistical simulations (the Monte Carlo method) is the generation of pseudo-random numbers that are distributed in a given way. Most algorithms are based on a pseudorandom number generator that produces numbers X that are uniformly distributed in the interval [0,1]. These random variates X are then transformed via some algorithm to create a new random variate having the required probability distribution.

Applications

The concept of the probability distribution and the random variables which they describe underlies the mathematical discipline of probability theory, and the science of statistics. There is spread or variability in almost any value that can be measured in a population (e.g. height of people, durability of a metal, sales growth, traffic flow, etc.); almost all measurements are made with some intrinsic error; in physics many processes are described probabilistically, from the kinetic properties of gases to the quantum mechanical description of fundamental particles. For these and many other reasons, simple numbers are often inadequate for describing a quantity, while probability distributions are often more appropriate.

As a more specific example of an application, the cache language models and other statistical language models used in natural language processing to assign probabilities to the occurrence of particular words and word sequences do so by means of probability distributions.

Common Probability Distributions

The following is a list of some of the most common probability distributions, grouped by the type of process that they are related to. For a more complete list, see list of probability distributions, which groups by the nature of the outcome being considered (discrete, continuous, multivariate, etc.) Note also that all of the univariate distributions below are singly peaked; that is, it is assumed that the values cluster around a single point. In practice, actually observed quantities may cluster around multiple values. Such quantities can be modeled using a mixture distribution.

Related to real-valued quantities that grow linearly (e.g. errors, offsets)

- Normal distribution (Gaussian distribution), for a single such quantity; the most common continuous distribution

Related to positive real-valued quantities that grow exponentially (e.g. prices, incomes, populations)

- Log-normal distribution, for a single such quantity whose log is normally distributed

- Pareto distribution, for a single such quantity whose log is exponentially distributed; the prototypical power law distribution

Related to Real-Valued Quantities that are assumed to be Uniformly Distributed over a (Possibly Unknown) Region

- Discrete uniform distribution, for a finite set of values (e.g. the outcome of a fair die)
- Continuous uniform distribution, for continuously distributed values

Related to Bernoulli trials (yes/no events, with a Given Probability)

- Basic distributions:
    - Bernoulli distribution, for the outcome of a single Bernoulli trial (e.g. success/failure, yes/no)
    - Binomial distribution, for the number of "positive occurrences" (e.g. successes, yes votes, etc.) given a fixed total number of independent occurrences
    - Negative binomial distribution, for binomial-type observations but where the quantity of interest is the number of failures before a given number of successes occurs
    - Geometric distribution, for binomial-type observations but where the quantity of interest is the number of failures before the first success; a special case of the negative binomial distribution
- Related to sampling schemes over a finite population:
    - Hypergeometric distribution, for the number of positive occurrences (e.g. successes, yes votes, etc.) given a fixed number of total occurrences, using sampling without replacement
    - Beta-binomial distribution, for the number of positive occurrences (e.g. successes, yes votes, etc.) given a fixed number of total occurrences, sampling using a Polya urn scheme (in some sense, the "opposite" of sampling without replacement)

Related to Categorical Outcomes (events with K Possible Outcomes, with a given Probability for each Outcome)

- Categorical distribution, for a single categorical outcome (e.g. yes/no/maybe in a survey); a generalization of the Bernoulli distribution
- Multinomial distribution, for the number of each type of categorical outcome, given a fixed number of total outcomes; a generalization of the binomial distribution

- Multivariate hypergeometric distribution, similar to the multinomial distribution, but using sampling without replacement; a generalization of the hypergeometric distribution

Related to events in a Poisson Process (events that occur independently with a given rate)

- Poisson distribution, for the number of occurrences of a Poisson-type event in a given period of time
- Exponential distribution, for the time before the next Poisson-type event occurs
- Gamma distribution, for the time before the next k Poisson-type events occur

Related to the Absolute Values of Vectors with Normally Distributed Components

- Rayleigh distribution, for the distribution of vector magnitudes with Gaussian distributed orthogonal components. Rayleigh distributions are found in RF signals with Gaussian real and imaginary components.
- Rice distribution, a generalization of the Rayleigh distributions for where there is a stationary background signal component. Found in Rician fading of radio signals due to multipath propagation and in MR images with noise corruption on non-zero NMR signals.

Related to Normally Distributed Quantities Operated with Sum of Squares (for Hypothesis Testing)

- Chi-squared distribution, the distribution of a sum of squared standard normal variables; useful e.g. for inference regarding the sample variance of normally distributed samples (see chi-squared test)
- Student›s t distribution, the distribution of the ratio of a standard normal variable and the square root of a scaled chi squared variable; useful for inference regarding the mean of normally distributed samples with unknown variance (see Student›s t-test)
- F-distribution, the distribution of the ratio of two scaled chi squared variables; useful e.g. for inferences that involve comparing variances or involving R-squared (the squared correlation coefficient)

Useful as Conjugate Prior Distributions in Bayesian Inference

- Beta distribution, for a single probability (real number between 0 and 1); conjugate to the Bernoulli distribution and binomial distribution

- Gamma distribution, for a non-negative scaling parameter; conjugate to the rate parameter of a Poisson distribution or exponential distribution, the precision (inverse variance) of a normal distribution, etc.
- Dirichlet distribution, for a vector of probabilities that must sum to 1; conjugate to the categorical distribution and multinomial distribution; generalization of the beta distribution
- Wishart distribution, for a symmetric non-negative definite matrix; conjugate to the inverse of the covariance matrix of a multivariate normal distribution; generalization of the gamma distribution

BETA DISTRIBUTION

In probability theory and statistics, the **beta distribution** is a family of continuous probability distributions defined on the interval [0, 1] parametrized by two positive shape parameters, denoted by α and β, that appear as exponents of the random variable and control the shape of the distribution.

The beta distribution has been applied to model the behavior of random variables limited to intervals of finite length in a wide variety of disciplines. For example, it has been used as a statistical description of allele frequencies in population genetics; time allocation in project management / control systems; sunshine data; variability of soil properties; proportions of the minerals in rocks in stratigraphy; and heterogeneity in the probability of HIV transmission.

In Bayesian inference, the beta distribution is the conjugate prior probability distribution for the Bernoulli, binomial, negative binomial and geometric distributions. For example, the beta distribution can be used in Bayesian analysis to describe initial knowledge concerning probability of success such as the probability that a space vehicle will successfully complete a specified mission. The beta distribution is a suitable model for the random behavior of percentages and proportions.

The usual formulation of the beta distribution is also known as the **beta distribution of the first kind**, whereas beta distribution of the second kind is an alternative name for the beta prime distribution.

Characterization

Probability Density Function

The probability density function (pdf) of the beta distribution, for $0 \leq x \leq 1$, and shape parameters $\alpha, \beta > 0$, is a power function of the variable x and of its reflection $(1 - x)$ as follows:

$$
\begin{aligned}
f(x;\alpha,\beta) &= \text{constant}\cdot x^{\alpha-1}(1-x)^{\beta-1} \\
&= \frac{x^{\alpha-1}(1-x)^{\beta-1}}{\int_0^1 u^{\alpha-1}(1-u)^{\beta-1}\,du} \\
&= \frac{\Gamma(\alpha+\beta)}{\Gamma(\alpha)\Gamma(\beta)}\,x^{\alpha-1}(1-x)^{\beta-1} \\
&= \frac{1}{\mathrm{B}(\alpha,\beta)}x^{\alpha-1}(1-x)^{\beta-1}
\end{aligned}
$$

where Γ(z) is the gamma function. The beta function, B, is a normalization constant to ensure that the total probability integrates to 1. In the above equations x is a realization—an observed value that actually occurred—of a random process X.

This definition includes both ends x = 0 and x = 1, which is consistent with definitions for other continuous distributions supported on a bounded interval which are special cases of the beta distribution, for example the arcsine distribution, and consistent with several authors, like N. L. Johnson and S. Kotz. However, the inclusion of x = 0 and x = 1 does not work for α, β < 1; accordingly, several other authors, including W. Feller, choose to exclude the ends x = 0 and x = 1, (such that the two ends are not actually part of the density function) and consider instead 0 < x < 1.

Several authors, including N. L. Johnson and S. Kotz, use the symbols p and q (instead of α and β) for the shape parameters of the beta distribution, reminiscent of the symbols traditionally used for the parameters of the Bernoulli distribution, because the beta distribution approaches the Bernoulli distribution in the limit when both shape parameters α and β approach the value of zero.

In the following, a random variable X beta-distributed with parameters α and β will be denoted by:

$$X \sim \mathrm{Beta}(\alpha,\beta)$$

Other notations for beta-distributed random variables used in the statistical literature are $X \sim \mathcal{B}e(\alpha,\beta)$ and $X \sim \beta_{\alpha,\beta}$.

Differential Equation

The probability density function satisfies the differential equation

$$f'(x) = f(x)\frac{(\alpha+\beta-2)x-(\alpha-1)}{(x-1)x}.$$

Cumulative Distribution Function

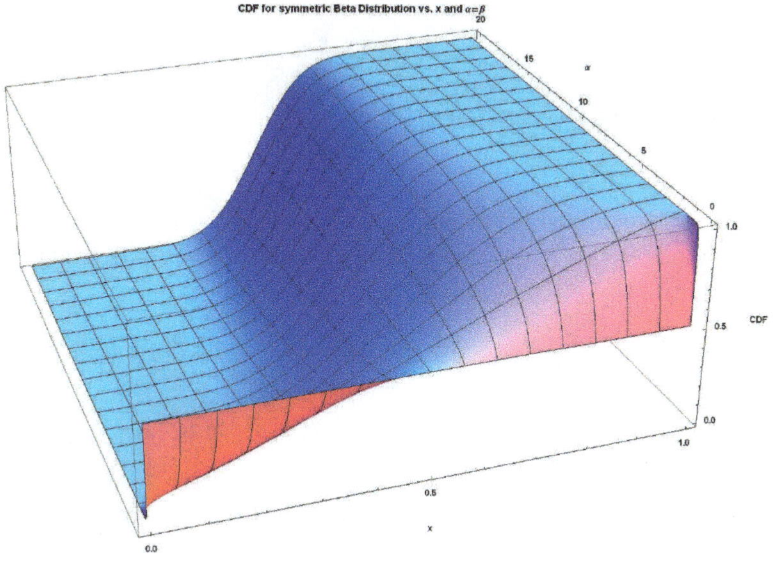

CDF for symmetric beta distribution vs. x and alpha=beta.

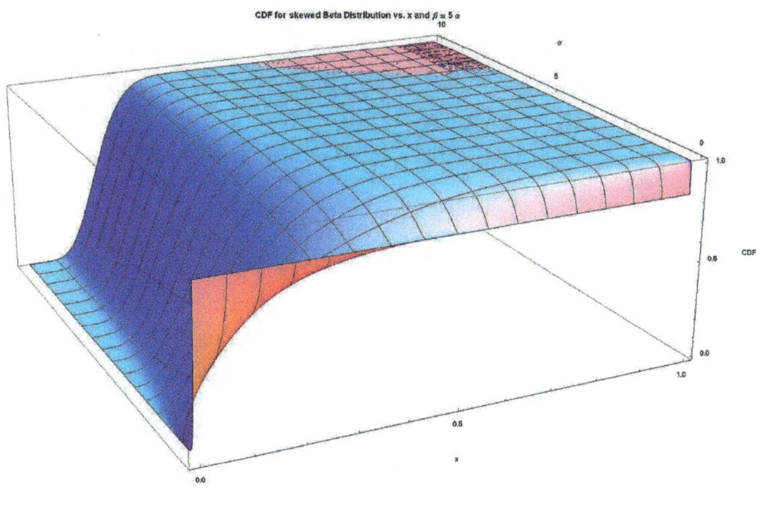

CDF for skewed beta distribution vs. x and beta= 5 alpha.

The cumulative distribution function is

$$F(x;\alpha,\beta) = \frac{\mathrm{B}(x;\alpha,\beta)}{\mathrm{B}(\alpha,\beta)} = I_x(\alpha,\beta)$$

where $\mathrm{B}(x;\alpha,\beta)$ is the incomplete beta function and $I_x(\alpha,\beta)$ is the regularized incomplete beta function.

Properties

Measures of Central Tendency

Mode

The mode of a Beta distributed random variable X with α, β > 1 is the most likely value of the distribution (corresponding to the peak in the PDF), and is given by the following expression:

$$\frac{\alpha - 1}{\alpha + \beta - 2}.$$

When both parameters are less than one (α, β < 1), this is the anti-mode: the lowest point of the probability density curve.

Letting α = β, the expression for the mode simplifies to 1/2, showing that for α = β > 1 the mode (resp. anti-mode when α, β < 1), is at the center of the distribution: it is symmetric in those cases. See Shapes section in this article for a full list of mode cases, for arbitrary values of α and β. For several of these cases, the maximum value of the density function occurs at one or both ends. In some cases the (maximum) value of the density function occurring at the end is finite. For example, in the case of α = 2, β = 1 (or α = 1, β = 2), the density function becomes a right-triangle distribution which is finite at both ends. In several other cases there is a singularity at one end, where the value of the density function approaches infinity. For example, in the case α = β = 1/2, the Beta distribution simplifies to become the arcsine distribution. There is debate among mathematicians about some of these cases and whether the ends (x = 0, and x = 1) can be called modes or not.

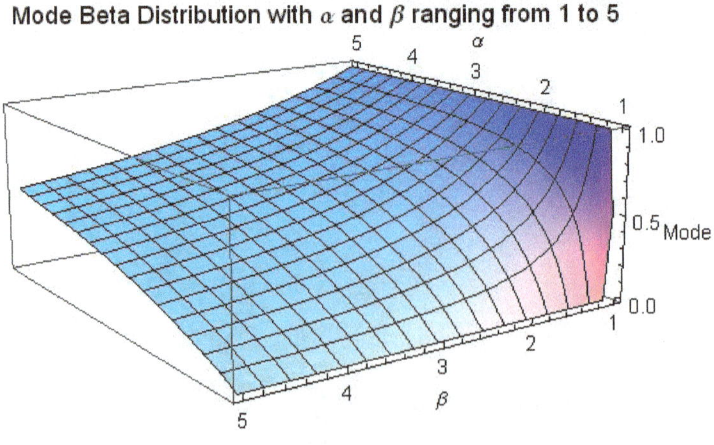

Mode for Beta distribution for $1 \leq \alpha \leq 5$ and $1 \leq \beta \leq 5$

- Whether the ends are part of the domain of the density function
- Whether a singularity can ever be called a mode
- Whether cases with two maxima should be called bimodal

Median

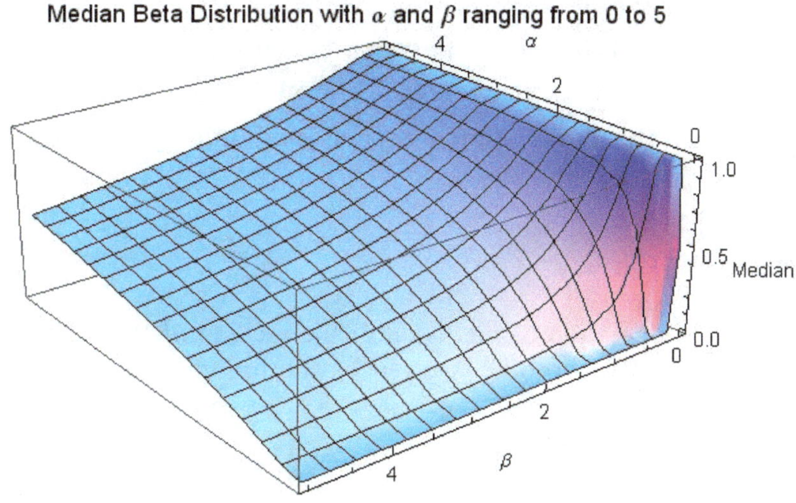

Median for Beta distribution for 0≤α≤5 and 0≤β≤5.

(Mean - Median) for Beta distribution versus alpha and beta from 0 to 2.

The median of the beta distribution is the unique real number $x = I_{\frac{1}{2}}^{[-1]}(\alpha, \beta)$ for which the regularized incomplete beta function $I_x(\alpha, \beta) = \frac{1}{2}$. There is no general closed-form expression for the median of the beta distribution for arbitrary values of α and β. Closed-form expressions for particular values of the parameters α and β follow:

- For symmetric cases α = β, median = 1/2.
- For α = 1 and β > 0, median = $1-2^{-\frac{1}{\beta}}$ (this case is the mirror-image of the power function [0,1] distribution)
- For α > 0 and β = 1, median = $2^{-\frac{1}{\alpha}}$ (this case is the power function [0,1] distribution)
- For α = 3 and β = 2, median = 0.6142724318676105..., the real solution to the quartic equation $1-8x^3+6x^4 = 0$, which lies in [0,1].
- For α = 2 and β = 3, median = 0.38572756813238945... = 1−median(Beta(3, 2))

The following are the limits with one parameter finite (non zero) and the other approaching these limits:

$$\lim_{\beta\to 0} \text{median} = \lim_{\alpha\to\infty} \text{median} = 1,$$

$$\lim_{\alpha\to 0} \text{median} = \lim_{\beta\to\infty} \text{median} = 0.$$

A reasonable approximation of the value of the median of the beta distribution, for both α and β greater or equal to one, is given by the formula

$$\text{median} \approx \frac{\alpha - \frac{1}{3}}{\alpha + \beta - \frac{2}{3}} \text{ for } \alpha, \beta \geq 1.$$

When α, β ≥ 1, the relative error (the absolute error divided by the median) in this approximation is less than 4% and for both α ≥ 2 and β ≥ 2 it is less than 1%. The absolute error divided by the difference between the mean and the mode is similarly small:

Mean

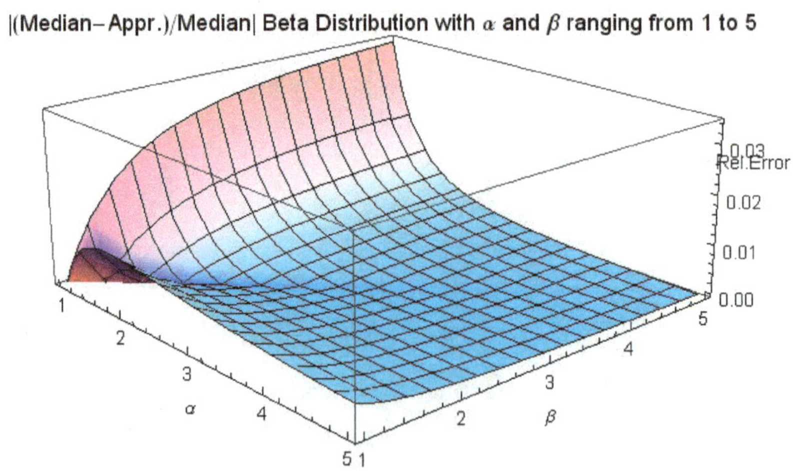

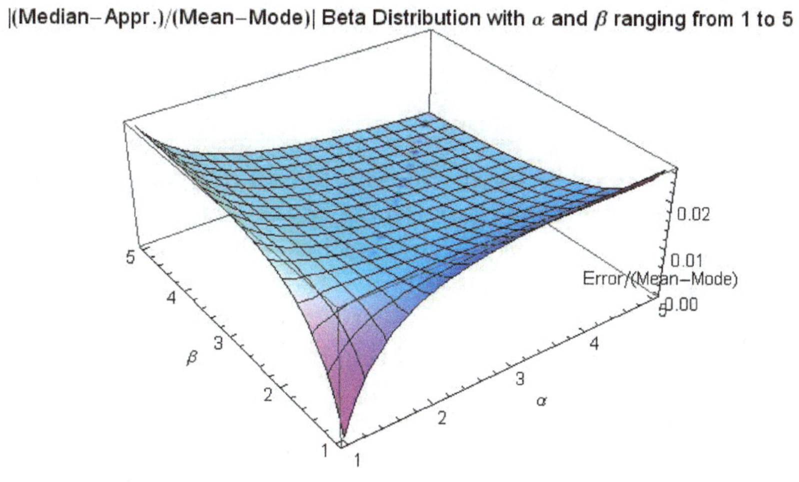

Mean for Beta distribution for $0 \leq \alpha \leq 5$ and $0 \leq \beta \leq 5$.

The expected value (mean) (μ) of a Beta distribution random variableX with two parameters α and β is a function of only the ratio β/α of these parameters:

$$\begin{aligned}\mu = \mathrm{E}[X] &= \int_0^1 x f(x;\alpha,\beta)\,dx \\ &= \int_0^1 x\,\frac{x^{\alpha-1}(1-x)^{\beta-1}}{\mathrm{B}(\alpha,\beta)}\,dx \\ &= \frac{\alpha}{\alpha+\beta} \\ &= \frac{1}{1+\frac{\beta}{\alpha}}\end{aligned}$$

Letting $\alpha = \beta$ in the above expression one obtains $\mu = 1/2$, showing that for $\alpha = \beta$ the mean is at the center of the distribution: it is symmetric. Also, the following limits can be obtained from the above expression:

$$\lim_{\frac{\beta}{\alpha}\to 0} \mu = 1$$

$$\lim_{\frac{\beta}{\alpha}\to \infty} \mu = 0$$

Therefore, for $\beta/\alpha \to 0$, or for $\alpha/\beta \to \infty$, the mean is located at the right end, $x = 1$. For these limit ratios, the beta distribution becomes a one-point degenerate distribution with a Dirac delta function spike at the right end, $x = 1$, with probability 1, and zero probability everywhere else. There is 100% probability (absolute certainty) concentrated at the right end, $x = 1$.

Similarly, for $\beta/\alpha \to \infty$, or for $\alpha/\beta \to 0$, the mean is located at the left end, $x = 0$. The beta distribution becomes a 1-point Degenerate distribution with a Dirac delta function spike at the left end, $x = 0$, with probability 1, and zero probability everywhere else. There is 100% probability (absolute certainty) concentrated at the left end, $x = 0$. Following are the limits with one parameter finite (non zero) and the other approaching these limits:

$$\lim_{\beta \to 0} \mu = \lim_{\alpha \to \infty} \mu = 1$$

$$\lim_{\alpha \to 0} \mu = \lim_{\beta \to \infty} \mu = 0$$

While for typical unimodal distributions (with centrally located modes, inflexion points at both sides of the mode, and longer tails) (with Beta(α, β) such that α, β > 2) it is known that the sample mean (as an estimate of location) is not as robust as the sample median, the opposite is the case for uniform or "U-shaped" bimodal distributions (with Beta(α, β) such that α, β ≤ 1), with the modes located at the ends of the distribution. As Mosteller and Tukey remark (p. 207) the average of the two extreme observations uses all the sample information. This illustrates how, for short-tailed distributions, the extreme observations should get more weight. By contrast, it follows that the median of U-shaped bimodal distributions with modes at the edge of the distribution (with Beta(α, β) such that α, β ≤ 1) is not robust, as the sample median drops the extreme sample observations from consideration. A practical application of this occurs for example for random walks, since the probability for the time of the last visit to the origin in a random walk is distributed as the arcsine distribution Beta(1/2, 1/2): the mean of a number of realizations of a random walk is a much more robust estimator than the median (which is an inappropriate sample measure estimate in this case).

Geometric mean

(Mean − GeometricMean) for Beta distribution versus α and β from 0 to 2, showing the asymmetry between α and β for the geometric mean

Geometric means for Beta distribution Purple = G(x), Yellow = G(1 − x), smaller values alpha and beta in front

Geometric means for Beta distribution. purple = G(x), yellow = G(1 − x), larger values α and β in front

The logarithm of the geometric mean G_X of a distribution with random variable X is the arithmetic mean of ln(X), or, equivalently, its expected value:

$$\ln G_X = \mathrm{E}[\ln X]$$

For a beta distribution, the expected value integral gives:

$$\begin{aligned}
E[\ln X] &= \int_0^1 \ln x \, f(x;\alpha,\beta)\, dx \\
&= \int_0^1 \ln x \frac{x^{\alpha-1}(1-x)^{\beta-1}}{B(\alpha,\beta)}\, dx \\
&= \frac{1}{B(\alpha,\beta)} \int_0^1 \frac{\partial x^{\alpha-1}(1-x)^{\beta-1}}{\partial \alpha}\, dx \\
&= \frac{1}{B(\alpha,\beta)} \frac{\partial}{\partial \alpha} \int_0^1 x^{\alpha-1}(1-x)^{\beta-1}\, dx \\
&= \frac{1}{B(\alpha,\beta)} \frac{\partial B(\alpha,\beta)}{\partial \alpha} \\
&= \frac{\partial \ln B(\alpha,\beta)}{\partial \alpha} \\
&= \frac{\partial \ln \Gamma(\alpha)}{\partial \alpha} - \frac{\partial \ln \Gamma(\alpha+\beta)}{\partial \alpha} \\
&= \psi(\alpha) - \psi(\alpha+\beta)
\end{aligned}$$

where ψ is the digamma function.

Therefore, the geometric mean of a beta distribution with shape parameters α and β is the exponential of the digamma functions of α and β as follows:

$$G_X = e^{E[\ln X]} = e^{\psi(\alpha)-\psi(\alpha+\beta)}$$

While for a beta distribution with equal shape parameters $\alpha = \beta$, it follows that skewness = 0 and mode = mean = median = 1/2, the geometric mean is less than 1/2: $0 < G_X < 1/2$. The reason for this is that the logarithmic transformation strongly weights the values of X close to zero, as ln(X) strongly tends towards negative infinity as X approaches zero, while ln(X) flattens towards zero as $X \to 1$.

Along a line $\alpha = \beta$, the following limits apply:

$$\lim_{\alpha=\beta\to 0} G_X = 0$$

$$\lim_{\alpha=\beta\to\infty} G_X = \tfrac{1}{2}$$

Following are the limits with one parameter finite (non zero) and the other approaching these limits:

$$\lim_{\beta\to 0} G_X = \lim_{\alpha\to\infty} G_X = 1$$

$$\lim_{\alpha\to 0} G_X = \lim_{\beta\to\infty} G_X = 0$$

The accompanying plot shows the difference between the mean and the geometric mean for shape parameters α and β from zero to 2. Besides the fact that the difference between them approaches zero as α and β approach infinity and that the difference becomes large for values of α and β approaching zero, one can observe an evident asymmetry of the geometric mean with respect to the shape parameters α and β. The difference between the geometric mean and the mean is larger for small values of α in relation to β than when exchanging the magnitudes of β and α.

N. L.Johnson and S. Kotz suggest the logarithmic approximation to the digamma function $\psi(\alpha) \approx \ln(\alpha\text{-}1/2)$ which results in the following approximation to the geometric mean:

$$G_X \approx \frac{\alpha - \frac{1}{2}}{\alpha + \beta - \frac{1}{2}} \text{ if } \alpha, \beta > 1.$$

Numerical values for the relative error in this approximation follow: [(α = β = 1): 9.39%]; [(α = β = 2): 1.29%]; [(α = 2, β = 3): 1.51%]; [(α = 3, β = 2): 0.44%]; [(α = β = 3): 0.51%]; [(α = β = 4): 0.26%]; [(α = 3, β = 4): 0.55%]; [(α = 4, β = 3): 0.24%].

Similarly, one can calculate the value of shape parameters required for the geometric mean to equal 1/2. Given the value of the parameter β, what would be the value of the other parameter, α, required for the geometric mean to equal 1/2?. The answer is that (for β > 1), the value of α required tends towards β + 1/2 as β → ∞. For example, all these couples have the same geometric mean of 1/2: [β = 1, α = 1.4427], [β = 2, α = 2.46958], [β = 3, α = 3.47943], [β = 4, α = 4.48449], [β = 5, α = 5.48756], [β = 10, α = 10.4938], [β = 100, α = 100.499].

The fundamental property of the geometric mean, which can be proven to be false for any other mean, is

$$G\left(\frac{X_i}{Y_i}\right) = \frac{G(X_i)}{G(Y_i)}$$

This makes the geometric mean the only correct mean when averaging normalized results, that is results that are presented as ratios to reference values. This is relevant because the beta distribution is a suitable model for the random behavior of percentages and it is particularly suitable to the statistical modelling of proportions. The geometric mean plays a central role in maximum likelihood estimation, see section Parameter estimation, maximum likelihood. Actually, when performing maximum likelihood estimation, besides the geometric mean G_X based on the random variable X, also another geometric mean appears naturally: the geometric mean based on the linear transformation —(1 − X), the mirror-image of X, denoted by $G_{(1-X)}$:

$$G_{(1-X)} = e^{\mathrm{E}[\ln(1-X)]} = e^{\psi(\beta)-\psi(\alpha+\beta)}$$

Along a line α = β, the following limits apply:

$$\lim_{\alpha=\beta\to 0} G_{(1-X)} = 0$$

$$\lim_{\alpha=\beta\to \infty} G_{(1-X)} = \frac{1}{2}$$

Following are the limits with one parameter finite (non-zero) and the other approaching these limits:

$$\lim_{\beta\to 0} G_{(1-X)} = \lim_{\alpha\to\infty} G_{(1-X)} = 0$$
$$\lim_{\alpha\to 0} G_{(1-X)} = \lim_{\beta\to\infty} G_{(1-X)} = 1$$

It has the following approximate value:

$$G_{(1-X)} \approx \frac{\beta - \frac{1}{2}}{\alpha + \beta - \frac{1}{2}} \text{ if } \alpha, \beta > 1.$$

Although both G_X and $G_{(1-X)}$ are asymmetric, in the case that both shape parameters are equal $\alpha = \beta$, the geometric means are equal: $G_X = G_{(1-X)}$. This equality follows from the following symmetry displayed between both geometric means:

$$G_X(\mathrm{B}(\alpha,\beta)) = G_{(1-X)}(\mathrm{B}(\beta,\alpha)).$$

Harmonic Mean

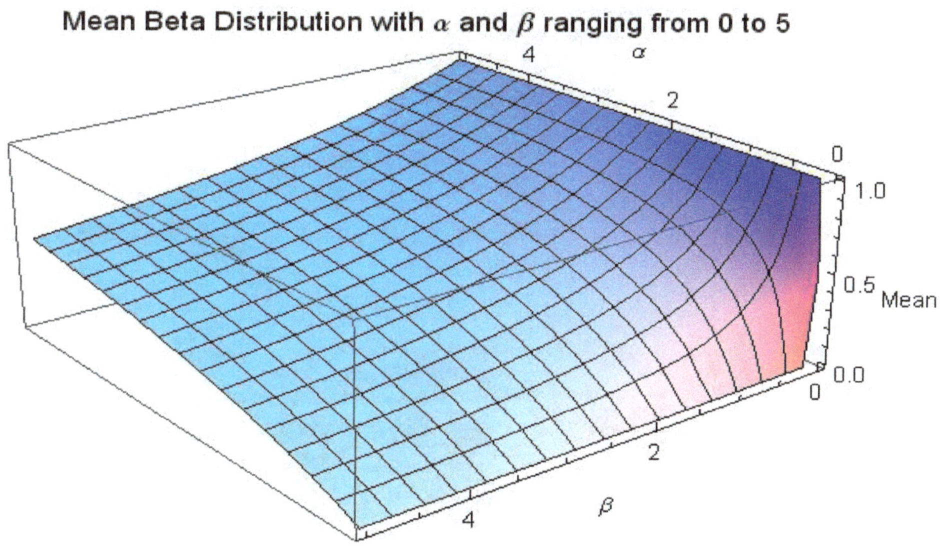

Harmonic mean for Beta distribution for 0<α<5 and 0<β<5.

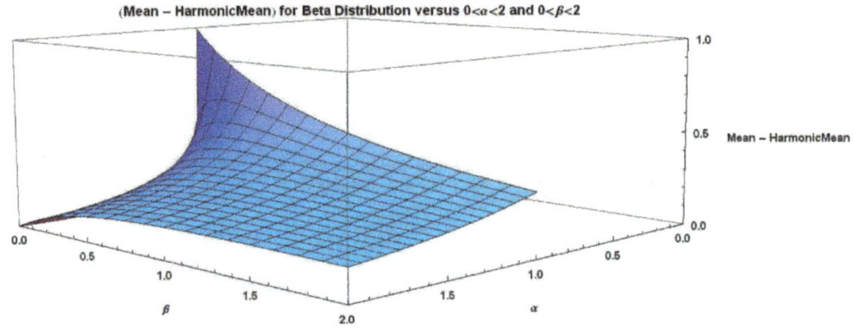

(Mean - HarmonicMean) for Beta distribution versus alpha and beta from 0 to 2.

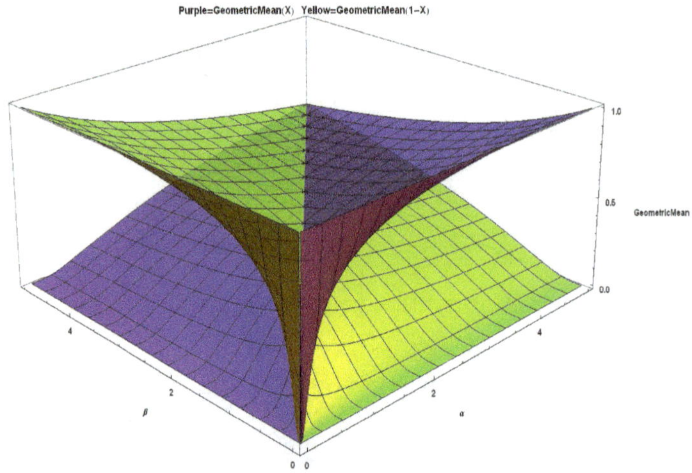

Harmonic Means for Beta distribution Purple=H(X), Yellow=H(1-X), smaller values alpha and beta in front.

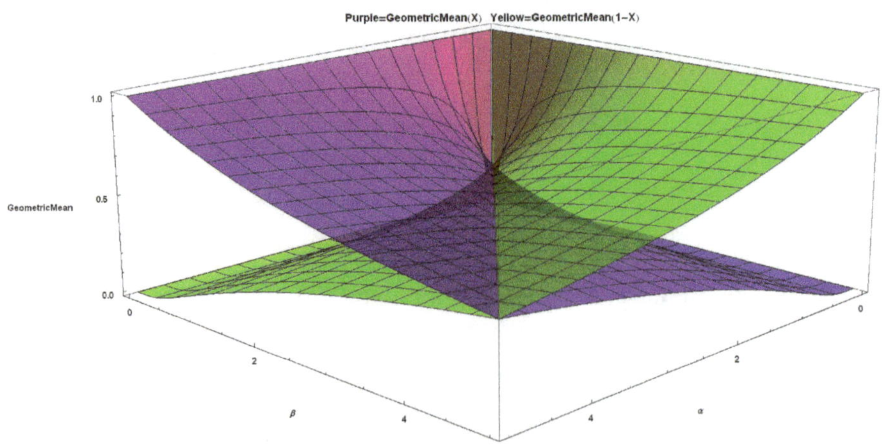

Harmonic Means for Beta distribution Purple=H(X), Yellow=H(1-X), larger values alpha and beta in front

The inverse of the harmonic mean (H_X) of a distribution with random variable X is the arithmetic mean of 1/X, or, equivalently, its expected value. Therefore, the harmonic mean (H_X) of a beta distribution with shape parameters α and β is:

$$\begin{aligned} H_X &= \frac{1}{\mathrm{E}\left[\frac{1}{X}\right]} \\ &= \frac{1}{\int_0^1 \frac{f(x;\alpha,\beta)}{x}\,dx} \\ &= \frac{1}{\int_0^1 \frac{x^{\alpha-1}(1-x)^{\beta-1}}{x\mathrm{B}(\alpha,\beta)}\,dx} \\ &= \frac{\alpha-1}{\alpha+\beta-1} \text{ if } \alpha > 1 \text{ and } \beta > 0 \end{aligned}$$

The harmonic mean (H_X) of a Beta distribution with α < 1 is undefined, because its defining expression is not bounded in [0, 1] for shape parameter α less than unity.

Letting α = β in the above expression one obtains

$$G_X = e^{\mathrm{E}[\ln X]} = e^{\psi(\alpha)-\psi(\alpha+\beta)}$$

showing that for α = β the harmonic mean ranges from 0, for α = β = 1, to 1/2, for α = β → ∞.

Following are the limits with one parameter finite (non zero) and the other approaching these limits:

$$\lim_{\alpha=\beta\to 0} G_X = 0$$

$$\lim_{\alpha=\beta\to\infty} G_X = \tfrac{1}{2}$$

The harmonic mean plays a role in maximum likelihood estimation for the four parameter case, in addition to the geometric mean. Actually, when performing maximum likelihood estimation for the four parameter case, besides the harmonic mean H_X based on the random variable X, also another harmonic mean appears naturally: the harmonic mean based on the linear transformation (1−X), the mirror-image of X, denoted by $H_{(1-X)}$:

$$G_{(1-X)} = e^{\mathrm{E}[\ln(1-X)]} = e^{\psi(\beta)-\psi(\alpha+\beta)}$$

Along a line $\alpha = \beta$, the following limits apply:

$$\lim_{\alpha=\beta\to 0} G_{(1-X)} = 0$$

$$\lim_{\alpha=\beta\to\infty} G_{(1-X)} = \tfrac{1}{2}$$

Following are the limits with one parameter finite (non-zero) and the other approaching these limits:

$$\lim_{\beta\to 0} G_{(1-X)} = \lim_{\alpha\to\infty} G_{(1-X)} = 0$$

$$\lim_{\alpha\to 0} G_{(1-X)} = \lim_{\beta\to\infty} G_{(1-X)} = 1$$

It has the following approximate value:

$$G_{(1-X)} \approx \frac{\beta-\frac{1}{2}}{\alpha+\beta-\frac{1}{2}} \text{ if } \alpha, \beta > 1.$$

Although both H_X and $H_{(1-X)}$ are asymmetric, in the case that both shape parameters are equal α = β, the harmonic means are equal: $H_X = H_{(1-X)}$. This equality follows from the following symmetry displayed between both harmonic means:

$$G_X(\mathrm{B}(\alpha,\beta)) = G_{(1-X)}(\mathrm{B}(\beta,\alpha)).$$

Measures of Statistical Dispersion

Variance

The variance (the second moment centered on the mean) of a Beta distribution random variable X with parameters α and β is:

$$\operatorname{var}(X) = \operatorname{E}[(X-\mu)^2] = \frac{\alpha\beta}{(\alpha+\beta)^2(\alpha+\beta+1)}$$

Letting α = β in the above expression one obtains

$$\operatorname{var}(X) = \frac{1}{4(2\beta+1)},$$

showing that for α = β the variance decreases monotonically as Template:Nowwrap increases. Setting α = β = 0 in this expression, one finds the maximum variance var(X) = 1/4 which only occurs approaching the limit, at α = β = 0.

The beta distribution may also be parametrized in terms of its mean μ (0 < μ < 1) and sample size ν = α + β (ν > 0) (see section below titled Mean and sample size):

$$\alpha = \mu\nu, \text{ where } \nu = (\alpha+\beta) > 0$$
$$\beta = (1-\mu)\nu, \text{ where } \nu = (\alpha+\beta) > 0.$$

Using this parametrization, one can express the variance in terms of the mean μ and the sample size ν as follows:

$$\operatorname{var}(X) = \frac{\mu(1-\mu)}{1+\nu}$$

Since {{{1}}}, it must follow that var(X) < μ(1 − μ)

For a symmetric distribution, the mean is at the middle of the distribution, {{{1}}}, and therefore:

$$\operatorname{var}(X) = \frac{1}{4(1+\nu)} \text{ if } \mu = \tfrac{1}{2}$$

Also, the following limits (with only the noted variable approaching the limit) can be obtained from the above expressions:

$$\lim_{\beta\to 0}\operatorname{var}(X) = \lim_{\alpha\to 0}\operatorname{var}(X) = \lim_{\beta\to\infty}\operatorname{var}(X) = \lim_{\alpha\to\infty}\operatorname{var}(X) = \lim_{\nu\to\infty}\operatorname{var}(X) = \lim_{\mu\to 0}\operatorname{var}(X) = \lim_{\mu\to 1}\operatorname{var}(X) = 0$$
$$\lim_{\nu\to 0}\operatorname{var}(X) = \mu(1-\mu)$$

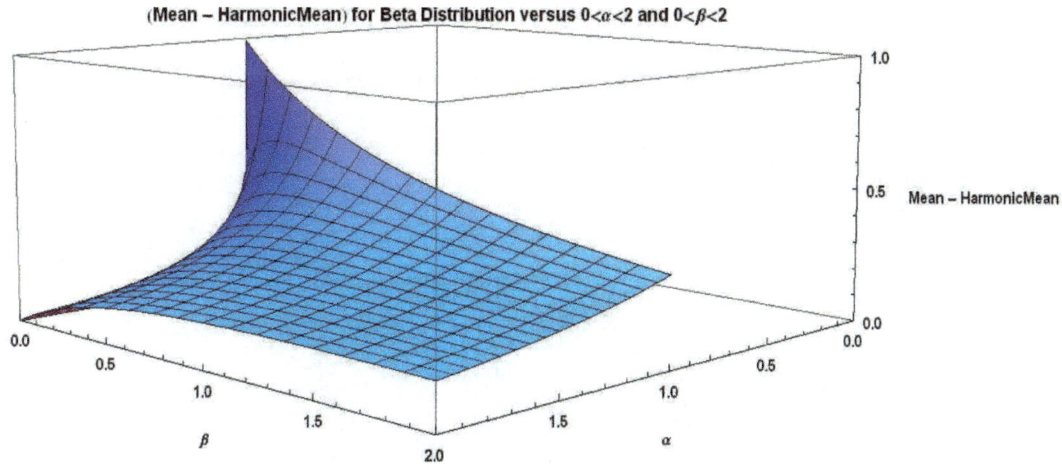

Geometric Variance and Covariance

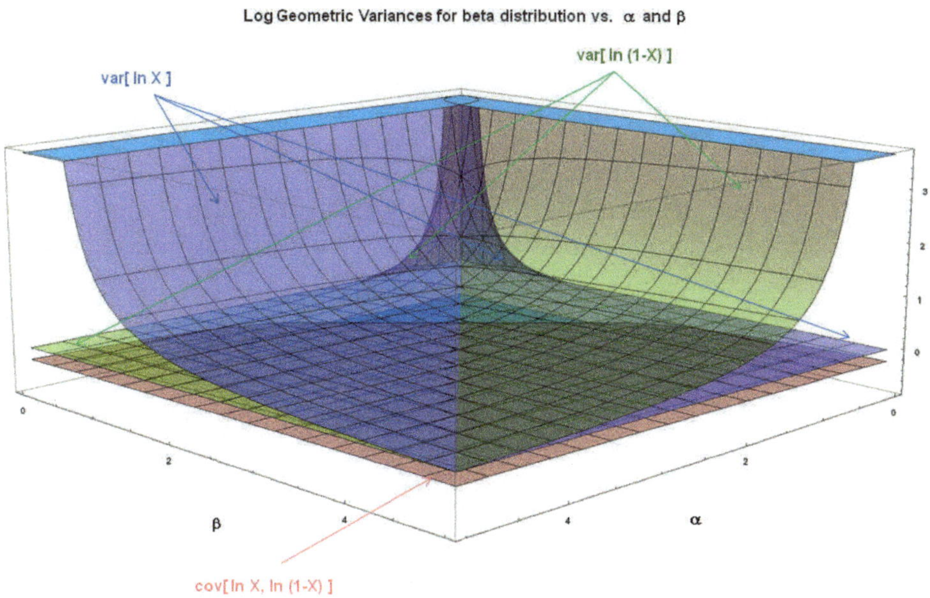

log geometric variances vs. α and β

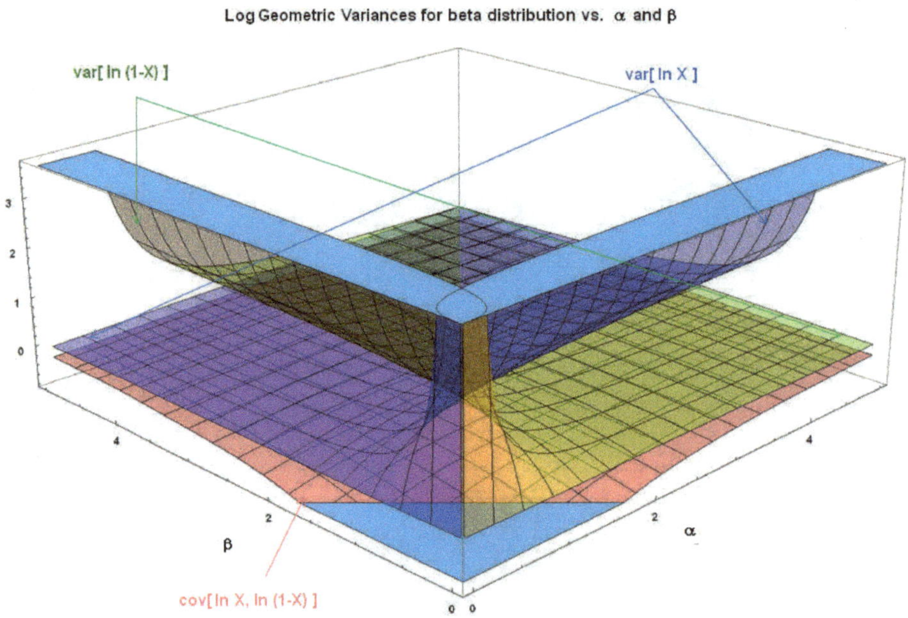

log geometric variances vs. α and β

The logarithm of the geometric variance, $\ln(\text{var}_{GX})$, of a distribution with random variable X is the second moment of the logarithm of X centered on the geometric mean of X, $(\ln(G_X)$:

$$\begin{aligned}\ln \text{var}_{GX} &= \mathrm{E}\left[(\ln X - \ln G_X)^2\right] \\ &= \mathrm{E}[(\ln X - \mathrm{E}[\ln X])^2] \\ &= \mathrm{E}\left[(\ln X)^2\right] - (\mathrm{E}[\ln X])^2 \\ &= \text{var}[\ln X]\end{aligned}$$

and therefore, the geometric variance is:

$$\text{var}_{GX} = e^{\text{var}[\ln X]}$$

In the Fisher information matrix, and the curvature of the log likelihood function, the logarithm of the geometric variance of the reflected variable (1-X) and the logarithm of the geometric covariance between X and (1-X) appear:

$$\begin{aligned}\ln \mathrm{var}_{G(1\text{-}X)} &= E[(\ln(1-X) - \ln G_{(1-X)})^2] \\ &= E[(\ln(1-X) - E[\ln(1-X)])^2] \\ &= E[(\ln(1-X))^2] - (E[\ln(1-X)])^2 \\ &= \mathrm{var}[\ln(1-X)]\end{aligned}$$

$$\mathrm{var}_{G(1\text{-}X)} = e^{\mathrm{var}[\ln(1-X)]}$$

$$\begin{aligned}\ln \mathrm{cov}_{GX,(1\text{-}X)} &= E[(\ln X - \ln G_X)(\ln(1-X) - \ln G_{(1-X)})] \\ &= E[(\ln X - E[\ln X])(\ln(1-X) - E[\ln(1-X)])] \\ &= E[\ln X \ln(1-X)] - E[\ln X]\, E[\ln(1-X)] \\ &= \mathrm{cov}[\ln X, \ln(1-X)]\end{aligned}$$

$$\mathrm{cov}_{GX,(1-X)} = e^{\mathrm{cov}[\ln X, \ln(1-X)]}$$

For a beta distribution, higher order logarithmic moments can be derived by using the representation of a beta distribution as a proportion of two Gamma distributions and differentiating through the integral. They can be expressed in terms of higher order polygamma functions. See the section titled "Other moments, Moments of transformed random variables, Moments of logarithmically transformed random variables". The variance of the logarithmic variables and covariance of lnX and ln(1−X) are:

$$\begin{aligned}&\mathrm{var}[\ln X] = \psi_1(\alpha) - \psi_1(\alpha+\beta) \\ &\mathrm{var}[\ln(1-X)] = \psi_1(\beta) - \psi_1(\alpha+\beta) \\ &\mathrm{cov}[\ln X, \ln(1-X)] = -\psi_1(\alpha+\beta)\end{aligned}$$

where the **trigamma function**, denoted $\psi_1(\alpha)$, is the second of the polygamma functions, and is defined as the derivative of the digamma function:

$$\psi_1(\alpha) = \frac{d^2 \ln \Gamma(\alpha)}{d\alpha^2} = \frac{d\,\psi(\alpha)}{d\alpha}.$$

Therefore,

$$\begin{aligned}\ln \mathrm{var}_{GX} &= \mathrm{var}[\ln X] = \psi_1(\alpha) - \psi_1(\alpha+\beta) \\ \ln \mathrm{var}_{G(1\text{-}X)} &= \mathrm{var}[\ln(1-X)] = \psi_1(\beta) - \psi_1(\alpha+\beta) \\ \ln \mathrm{cov}_{GX,(1\text{-}X)} &= \mathrm{cov}[\ln X, \ln(1-X)] = -\psi_1(\alpha+\beta)\end{aligned}$$

The accompanying plots show the log geometric variances and log geometric covariance versus the shape parameters α and β. The plots show that the log geometric variances and log geometric covariance are close to zero for shape parameters α and β greater than 2, and that the log geometric variances rapidly rise in value for shape parameter values α and β less than unity. The log geometric variances are positive for all values of the shape parameters. The log geometric covariance is negative for all values of the shape parameters, and it reaches large negative values for α and β less than unity.

Following are the limits with one parameter finite (non zero) and the other approaching these limits:

$$\lim_{\alpha\to 0} \ln \text{var}_{GX} = \lim_{\beta\to 0} \ln \text{var}_{G(1-X)} = \infty$$

$$\lim_{\beta\to 0} \ln \text{var}_{GX} = \lim_{\alpha\to\infty} \ln \text{var}_{GX} = \lim_{\alpha\to 0} \ln \text{var}_{G(1-X)} = \lim_{\beta\to\infty} \ln \text{var}_{G(1-X)} = \lim_{\alpha\to\infty} \ln \text{cov}_{GX,(1-X)} = \lim_{\beta\to\infty} \ln \text{cov}_{GX,(1-X)} = 0$$

$$\lim_{\beta\to\infty} \ln \text{var}_{GX} = \psi_1(\alpha)$$

$$\lim_{\alpha\to\infty} \ln \text{var}_{G(1-X)} = \psi_1(\beta)$$

$$\lim_{\alpha\to 0} \ln \text{cov}_{GX,(1-X)} = -\psi_1(\beta)$$

$$\lim_{\beta\to 0} \ln \text{cov}_{GX,(1-X)} = -\psi_1(\alpha)$$

Limits with two parameters varying:

$$\lim_{\alpha\to\infty}\left(\lim_{\beta\to\infty} \ln \text{var}_{GX}\right) = \lim_{\beta\to\infty}\left(\lim_{\alpha\to\infty} \ln \text{var}_{G(1-X)}\right) = \lim_{\alpha\to\infty}\left(\lim_{\beta\to 0} \ln \text{cov}_{GX,(1-X)}\right) = \lim_{\beta\to\infty}\left(\lim_{\alpha\to 0} \ln \text{cov}_{GX,(1-X)}\right) = 0$$

$$\lim_{\alpha\to\infty}\left(\lim_{\beta\to 0} \ln \text{var}_{GX}\right) = \lim_{\beta\to\infty}\left(\lim_{\alpha\to 0} \ln \text{var}_{G(1-X)}\right) = \infty$$

$$\lim_{\alpha\to 0}\left(\lim_{\beta\to 0} \ln \text{cov}_{GX,(1-X)}\right) = \lim_{\beta\to 0}\left(\lim_{\alpha\to 0} \ln \text{cov}_{GX,(1-X)}\right) = -\infty$$

Although both $\ln(\text{var}_{GX})$ and $\ln(\text{var}_{G(1-X)})$ are asymmetric, when the shape parameters are equal, α = β, one has: $\ln(\text{var}_{GX}) = \ln(\text{var}_{G(1-X)})$. This equality follows from the following symmetry displayed between both log geometric variances:

$$\ln \text{var}_{GX}(\mathrm{B}(\alpha,\beta)) = \ln \text{var}_{G(1-X)}(\mathrm{B}(\beta,\alpha)).$$

The log geometric covariance is symmetric:

$$\ln \text{cov}_{GX,(1-X)}(\mathrm{B}(\alpha,\beta)) = \ln \text{cov}_{GX,(1-X)}(\mathrm{B}(\beta,\alpha))$$

Mean Absolute Deviation Around the Mean

Ratio of Mean Abs.Dev. to Std.Dev. for Beta distribution with α and β ranging from 0 to 5

Ratio of Mean Abs.Dev. to Std.Dev. for Beta distribution with mean $0 \leq \mu \leq 1$ and sample size $0 < \nu \leq 10$

The mean absolute deviation around the mean for the beta distribution with shape parameters α and β is:

$$\mathrm{E}[|X - E[X]|] = \frac{2\alpha^\alpha \beta^\beta}{\mathrm{B}(\alpha,\beta)(\alpha+\beta)^{\alpha+\beta+1}}$$

The mean absolute deviation around the mean is a more robust estimator of statistical dispersion than the standard deviation for beta distributions with tails and inflection points at each side of the mode, Beta(α, β) distributions with α,β > 2, as it depends on the linear (absolute) deviations rather than the square deviations from the mean. Therefore, the effect of very large deviations from the mean are not as overly weighted.

Using Stirling›s approximation to the Gamma function, N.L.Johnson and S.Kotzderived the following approximation for values of the shape parameters greater than unity (the relative error for this approximation is only −3.5% for α = β = 1, and it decreases to zero as α → ∞, β → ∞):

$$\frac{\text{mean abs. dev. from mean}}{\text{standard deviation}} = \frac{\mathrm{E}[|X - E[X]|]}{\sqrt{\mathrm{var}(X)}}$$
$$\approx \sqrt{\frac{2}{\pi}}\left(1 + \frac{7}{12(\alpha+\beta)} - \frac{1}{12\alpha} - \frac{1}{12\beta}\right), \text{ if } \alpha, \beta > 1.$$

At the limit $\alpha \to \infty$, $\beta \to \infty$, the ratio of the mean absolute deviation to the standard deviation (for the beta distribution) becomes equal to the ratio of the same measures for the normal distribution: $\sqrt{\frac{2}{\pi}}$. For $\alpha = \beta = 1$ this ratio equals $\frac{\sqrt{3}}{2}$, so that from $\alpha = \beta = 1$ to α, $\beta \to \infty$ the ratio decreases by 8.5%. For $\alpha = \beta = 0$ the standard deviation is exactly equal to the mean absolute deviation around the mean. Therefore, this ratio decreases by 15% from $\alpha = \beta = 0$ to $\alpha = \beta = 1$, and by 25% from $\alpha = \beta = 0$ to α, $\beta \to \infty$. However, for skewed beta distributions such that $\alpha \to 0$ or $\beta \to 0$, the ratio of the standard deviation to the mean absolute deviation approaches infinity (although each of them, individually, approaches zero) because the mean absolute deviation approaches zero faster than the standard deviation.

Using the parametrization in terms of mean μ and sample size $\nu = \alpha + \beta > 0$:
$\alpha = \mu\nu$, $\beta = (1-\mu)\nu$

one can express the mean absolute deviation around the mean in terms of the mean μ and the sample size ν as follows:

$$\mathrm{E}[|X - E[X]|] = \frac{2\mu^{\mu\nu}(1-\mu)^{(1-\mu)\nu}}{\nu \mathrm{B}(\mu\nu, (1-\mu)\nu)}$$

For a symmetric distribution, the mean is at the middle of the distribution, $\mu = 1/2$, and therefore:

$$\mathrm{E}[|X - E[X]|] = \frac{2^{1-\nu}}{\nu \mathrm{B}(\frac{\nu}{2}, \frac{\nu}{2})} = \frac{2^{1-\nu}\Gamma(\nu)}{\nu(\Gamma(\frac{\nu}{2}))^2}$$

$$\lim_{\nu \to 0}\left(\lim_{\mu \to \frac{1}{2}} \mathrm{E}[|X - E[X]|]\right) = \frac{1}{2}$$

$$\lim_{\nu \to \infty}\left(\lim_{\mu \to \frac{1}{2}} \mathrm{E}[|X - E[X]|]\right) = 0$$

Also, the following limits (with only the noted variable approaching the limit) can be obtained from the above expressions:

$$\lim_{\beta \to 0} \mathrm{E}[|X - E[X]|] = \lim_{\alpha \to 0} \mathrm{E}[|X - E[X]|] = 0$$
$$\lim_{\beta \to \infty} \mathrm{E}[|X - E[X]|] = \lim_{\alpha \to \infty} \mathrm{E}[|X - E[X]|] = 0$$
$$\lim_{\mu \to 0} \mathrm{E}[|X - E[X]|] = \lim_{\mu \to 1} \mathrm{E}[|X - E[X]|] = 0$$
$$\lim_{\nu \to 0} \mathrm{E}[|X - E[X]|] = \sqrt{\mu(1-\mu)}$$
$$\lim_{\nu \to \infty} \mathrm{E}[|X - E[X]|] = 0$$

Mean Absolute Difference

The mean absolute difference for the Beta distribution is:

$$\mathrm{MD} = \int_0^1 \int_0^1 f(x;\alpha,\beta)\, f(y;\alpha,\beta)\, |x-y|\, dx\, dy = \left(\frac{4}{\alpha+\beta}\right) \frac{B(\alpha+\beta,\alpha+\beta)}{B(\alpha,\alpha)B(\beta,\beta)}$$

The Gini coefficient for the Beta distribution is half of the relative mean absolute difference:

$$\mathrm{G} = \left(\frac{2}{\alpha}\right) \frac{B(\alpha+\beta,\alpha+\beta)}{B(\alpha,\alpha)B(\beta,\beta)}$$

Skewness

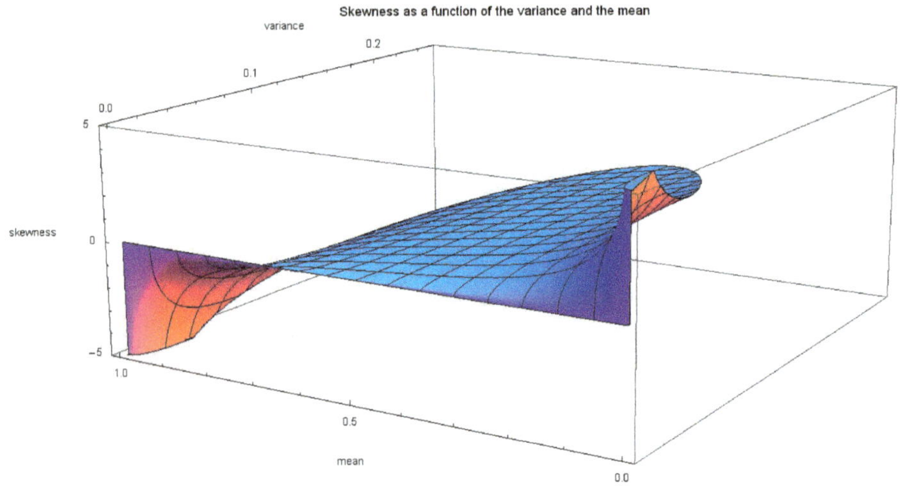

Skewness for Beta Distribution as a function of variance and mean.

The skewness (the third moment centered on the mean, normalized by the 3/2 power of the variance) of the beta distribution is

$$\gamma_1 = \frac{\mathrm{E}[(X-\mu)^3]}{(\mathrm{var}(X))^{3/2}} = \frac{2(\beta-\alpha)\sqrt{\alpha+\beta+1}}{(\alpha+\beta+2)\sqrt{\alpha\beta}}.$$

Letting $\alpha = \beta$ in the above expression one obtains $\gamma_1 = 0$, showing once again that for $\alpha = \beta$ the distribution is symmetric and hence the skewness is zero. Positive skew (right-tailed) for $\alpha < \beta$, negative skew (left-tailed) for $\alpha > \beta$.

Using the parametrization in terms of mean μ and sample size $\nu = \alpha + \beta$:

$$\alpha = \mu\nu, \text{ where } \nu = (\alpha + \beta) > 0$$
$$\beta = (1 - \mu)\nu, \text{ where } \nu = (\alpha + \beta) > 0.$$

one can express the skewness in terms of the mean μ and the sample size ν as follows:

$$\gamma_1 = \frac{\mathrm{E}[(X-\mu)^3]}{(\mathrm{var}(X))^{3/2}} = \frac{2(1-2\mu)\sqrt{1+\nu}}{(2+\nu)\sqrt{\mu(1-\mu)}}.$$

The skewness can also be expressed just in terms of the variance *var* and the mean μ as follows:

$$\gamma_1 = \frac{\mathrm{E}[(X-\mu)^3]}{(\mathrm{var}(X))^{3/2}} = \frac{2(1-2\mu)\sqrt{\mathrm{var}}}{\mu(1-\mu)+\mathrm{var}} \text{ if } \mathrm{var} < \mu(1-\mu)$$

The accompanying plot of skewness as a function of variance and mean shows that maximum variance (1/4) is coupled with zero skewness and the symmetry condition (μ = 1/2), and that maximum skewness (positive or negative infinity) occurs when the mean is located at one end or the other, so that the "mass" of the probability distribution is concentrated at the ends (minimum variance).

The following expression for the square of the skewness, in terms of the sample size ν = α + β and the variance var, is useful for the method of moments estimation of four parameters:

$$(\gamma_1)^2 = \frac{(\mathrm{E}[(X-\mu)^3])^2}{(\mathrm{var}(X))^3} = \frac{4}{(2+\nu)^2}\left(\frac{1}{\mathrm{var}} - 4(1+\nu)\right)$$

This expression correctly gives a skewness of zero for α = β, since in that case (see section titled "Variance"): $\mathrm{var} = \frac{1}{4(1+\nu)}$.

For the symmetric case (α = β), skewness = 0 over the whole range, and the following limits apply:

$$\lim_{\alpha=\beta\to 0}\gamma_1 = \lim_{\alpha=\beta\to\infty}\gamma_1 = \lim_{\nu\to 0}\gamma_1 = \lim_{\nu\to\infty}\gamma_1 = \lim_{\mu\to\frac{1}{2}}\gamma_1 = 0$$

For the asymmetric cases (α ≠ β) the following limits (with only the noted variable approaching the limit) can be obtained from the above expressions:

$$\lim_{\alpha\to 0}\gamma_1 = \lim_{\mu\to 0}\gamma_1 = \infty$$
$$\lim_{\beta\to 0}\gamma_1 = \lim_{\mu\to 1}\gamma_1 = -\infty$$
$$\lim_{\alpha\to\infty}\gamma_1 = -\frac{2}{\beta}, \quad \lim_{\beta\to 0}(\lim_{\alpha\to\infty}\gamma_1) = -\infty, \quad \lim_{\beta\to\infty}(\lim_{\alpha\to\infty}\gamma_1) = 0$$
$$\lim_{\beta\to\infty}\gamma_1 = \frac{2}{\alpha}, \quad \lim_{\alpha\to 0}(\lim_{\beta\to\infty}\gamma_1) = \infty, \quad \lim_{\alpha\to\infty}(\lim_{\beta\to\infty}\gamma_1) = 0$$
$$\lim_{\nu\to 0}\gamma_1 = \frac{1-2\mu}{\sqrt{\mu(1-\mu)}}, \quad \lim_{\mu\to 0}(\lim_{\nu\to 0}\gamma_1) = \infty, \quad \lim_{\mu\to 1}(\lim_{\nu\to 0}\gamma_1) = -\infty$$

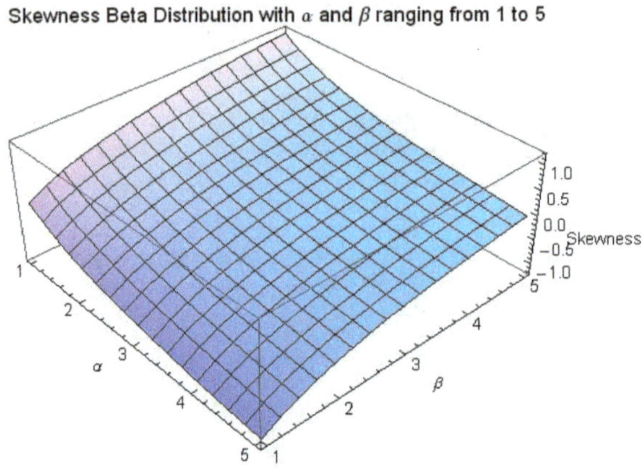

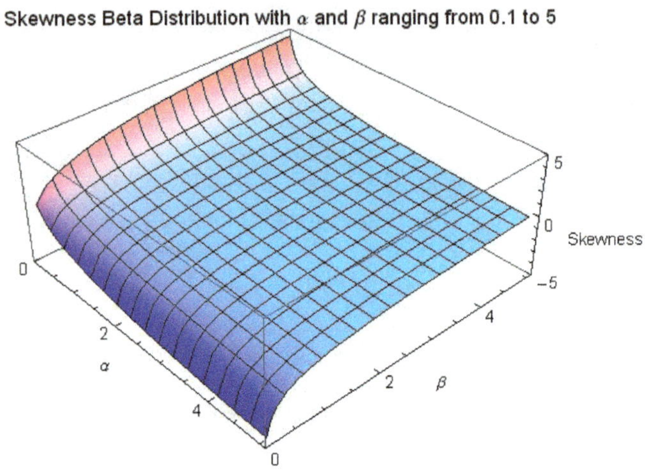

Kurtosis

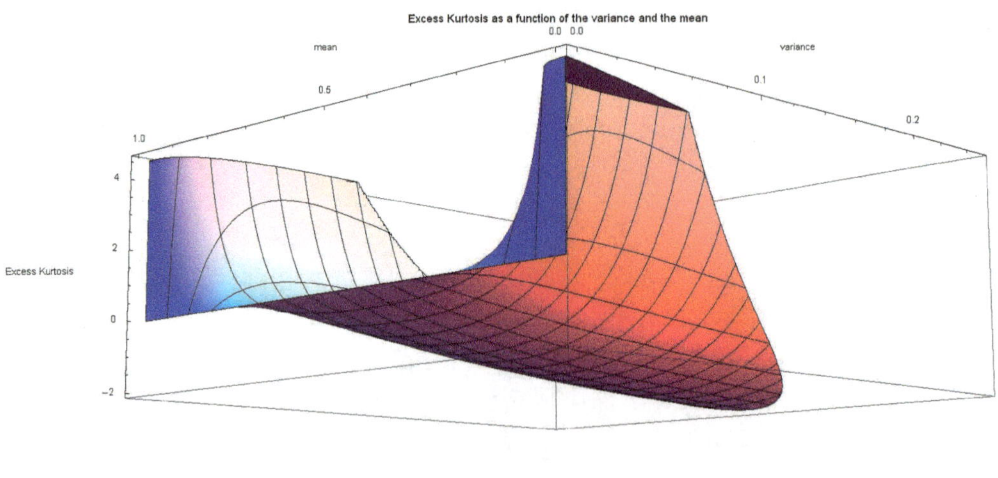

Excess Kurtosis for Beta Distribution as a function of variance and mean.

The beta distribution has been applied in acoustic analysis to assess damage to gears, as the kurtosis of the beta distribution has been reported to be a good indicator of the condition of a gear. Kurtosis has also been used to distinguish the seismic signal generated by a person's footsteps from other signals. As persons or other targets moving on the ground generate continuous signals in the form of seismic waves, one can separate different targets based on the seismic waves they generate. Kurtosis is sensitive to impulsive signals, so it's much more sensitive to the signal generated by human footsteps than other signals generated by vehicles, winds, noise, etc. Unfortunately, the notation for kurtosis has not been standardized. Kenney and Keeping use the symbol γ_2 for the excess kurtosis, but Abramowitz and Stegun use different terminology. To prevent confusion between kurtosis (the fourth moment centered on the mean, normalized by the square of the variance) and excess kurtosis, when using symbols, they will be spelled out as follows:

$$\begin{aligned}\text{excess kurtosis} &= \text{kurtosis} - 3\\ &= \frac{\mathrm{E}[(X-\mu)^4]}{(\mathrm{var}(X))^2} - 3\\ &= \frac{6[\alpha^3 - \alpha^2(2\beta - 1) + \beta^2(\beta + 1) - 2\alpha\beta(\beta + 2)]}{\alpha\beta(\alpha + \beta + 2)(\alpha + \beta + 3)}\\ &= \frac{6[(\alpha - \beta)^2(\alpha + \beta + 1) - \alpha\beta(\alpha + \beta + 2)]}{\alpha\beta(\alpha + \beta + 2)(\alpha + \beta + 3)}.\end{aligned}$$

Letting $\alpha = \beta$ in the above expression one obtains

$$\text{excess kurtosis} = -\frac{6}{3 + 2\alpha} \text{ if } \alpha = \beta.$$

Therefore, for symmetric beta distributions, the excess kurtosis is negative, increasing from a minimum value of −2 at the limit as $\{\alpha = \beta\} \to 0$, and approaching a maximum value of zero as $\{\alpha = \beta\} \to \infty$. The value of −2 is the minimum value of excess kurtosis that any distribution (not just beta distributions, but any distribution of any possible kind) can ever achieve. This minimum value is reached when all the probability density is entirely concentrated at each end $x = 0$ and $x = 1$, with nothing in between: a 2-point Bernoulli distribution with equal probability 1/2 at each end (a coin toss: see section below "Kurtosis bounded by the square of the skewness" for further discussion). The description of kurtosis as a measure of the "peakedness" (or "heavy tails") of the probability distribution, is strictly applicable to unimodal distributions (for example the normal distribution). However, for more general distributions, like the beta distribution, a more general description of kurtosis is that it is a measure of the proportion of the mass density near the mean. The higher the proportion of mass density near the mean, the higher the kurtosis, while the higher the mass density away from the mean, the lower the kurtosis. For $\alpha \neq \beta$, skewed beta distributions, the excess kurtosis can reach unlimited positive values (particularly for $\alpha \to 0$ for finite β, or for $\beta \to 0$ for finite α) because all the mass density is concentrated at the mean when the mean coincides with one of the ends. Minimum kurtosis takes place when the mass density is concentrated equally at each end (and therefore the mean is at the center), and there is no probability mass density in between the ends.

Using the parametrization in terms of mean μ and sample size $\nu = \alpha + \beta$:

$$\alpha = \mu\nu, \text{ where } \nu = (\alpha + \beta) > 0$$
$$\beta = (1 - \mu)\nu, \text{ where } \nu = (\alpha + \beta) > 0.$$

one can express the excess kurtosis in terms of the mean μ and the sample size ν as follows:

$$\text{excess kurtosis} = \frac{6}{3+\nu}\left(\frac{(1-2\mu)^2(1+\nu)}{\mu(1-\mu)(2+\nu)} - 1\right)$$

The excess kurtosis can also be expressed in terms of just the following two parameters: the variance var, and the sample size ν as follows:

$$\text{excess kurtosis} = \frac{6}{(3+\nu)(2+\nu)}\left(\frac{1}{\text{var}} - 6 - 5\nu\right) \text{ if var } < \mu(1-\mu)$$

and, in terms of the variance *var* and the mean μ as follows:

$$\text{excess kurtosis} = \frac{6 \text{ var } (1 - \text{ var } - 5\mu(1-\mu))}{(\text{var } + \mu(1-\mu))(2 \text{ var } + \mu(1-\mu))} \text{ if var } < \mu(1-\mu)$$

The plot of excess kurtosis as a function of the variance and the mean shows that the minimum value of the excess kurtosis (−2, which is the minimum possible value for excess kurtosis for any distribution) is intimately coupled with the maximum value of variance (1/4) and the symmetry condition: the mean occurring at the midpoint (μ = 1/2). This occurs for the symmetric case of α = β = 0, with zero skewness. At the limit, this is the 2 point Bernoulli distribution with equal probability 1/2 at each Dirac delta function end x = 0 and x = 1 and zero probability everywhere else. (A coin toss: one face of the coin being x = 0 and the other face being x = 1.) Variance is maximum because the distribution is bimodal with nothing in between the two modes (spikes) at each end. Excess kurtosis is minimum: the probability density "mass" is zero at the mean and it is concentrated at the two peaks at each end. Excess kurtosis reaches the minimum possible value (for any distribution) when the probability density function has two spikes at each end: it is bi-"peaky" with nothing in between them.

On the other hand, the plot shows that for extreme skewed cases, where the mean is located near one or the other end (μ = 0 or μ = 1), the variance is close to zero, and the excess kurtosis rapidly approaches infinity when the mean of the distribution approaches either end.

Alternatively, the excess kurtosis can also be expressed in terms of just the following two parameters: the square of the skewness, and the sample size ν as follows:

$$\text{excess kurtosis} = \frac{6}{3+\nu}\left(\frac{(2+\nu)}{4}(\text{skewness})^2 - 1\right) \text{ if } (\text{skewness})^2 - 2 < \text{excess kurtosis} < \frac{3}{2}(\text{skewness})^2$$

From this last expression, one can obtain the same limits published practically a century ago by Karl Pearson in his paper,for the beta distribution (see section below titled Kurtosis bounded by the square of the skewness). Setting α + β= ν = 0 in the above expression, one obtains Pearson›s lower boundary (values for the skewness and excess kurtosis below the boundary (excess kurtosis + 2 − skewness^2 = 0) cannot occur for any distribution, and hence Karl Pearson appropriately called the region below this boundary the “impossible region”). The limit of $\alpha + \beta = \nu \to \infty$ determines Pearson’s upper boundary.

$$\lim_{\nu \to 0} \text{excess kurtosis} = (\text{skewness})^2 - 2$$

$$\lim_{\nu \to \infty} \text{excess kurtosis} = \tfrac{3}{2}(\text{skewness})^2$$

therefore:

$$(\text{skewness})^2 - 2 < \text{excess kurtosis} < \tfrac{3}{2}(\text{skewness})^2$$

Values of $\nu = \alpha + \beta$ such that ν ranges from zero to infinity, $0 < \nu < \infty$, span the whole region of the beta distribution in the plane of excess kurtosis versus squared skewness.

For the symmetric case ($\alpha = \beta$), the following limits apply:

$$\lim_{\alpha=\beta \to 0} \text{excess kurtosis} = -2$$

$$\lim_{\alpha=\beta \to \infty} \text{excess kurtosis} = 0$$

$$\lim_{\mu \to \frac{1}{2}} \text{excess kurtosis} = -\frac{6}{3+\nu}$$

For the unsymmetric cases ($\alpha \neq \beta$) the following limits (with only the noted variable approaching the limit) can be obtained from the above expressions:

$$\lim_{\alpha \to 0} \text{excess kurtosis} = \lim_{\beta \to 0} \text{excess kurtosis} = \lim_{\mu \to 0} \text{excess kurtosis} = \lim_{\mu \to 1} \text{excess kurtosis} = \infty$$

$$\lim_{\alpha \to \infty} \text{excess kurtosis} = \frac{6}{\beta}, \quad \lim_{\beta \to 0}(\lim_{\alpha \to \infty} \text{excess kurtosis}) = \infty, \quad \lim_{\beta \to \infty}(\lim_{\alpha \to \infty} \text{excess kurtosis}) = 0$$

$$\lim_{\beta \to \infty} \text{excess kurtosis} = \frac{6}{\alpha}, \quad \lim_{\alpha \to 0}(\lim_{\beta \to \infty} \text{excess kurtosis}) = \infty, \quad \lim_{\alpha \to \infty}(\lim_{\beta \to \infty} \text{excess kurtosis}) = 0$$

$$\lim_{\nu \to 0} \text{excess kurtosis} = -6 + \frac{1}{\mu(1-\mu)}, \quad \lim_{\mu \to 0}(\lim_{\nu \to 0} \text{excess kurtosis}) = \infty, \quad \lim_{\mu \to 1}(\lim_{\nu \to 0} \text{excess kurtosis}) = \infty$$

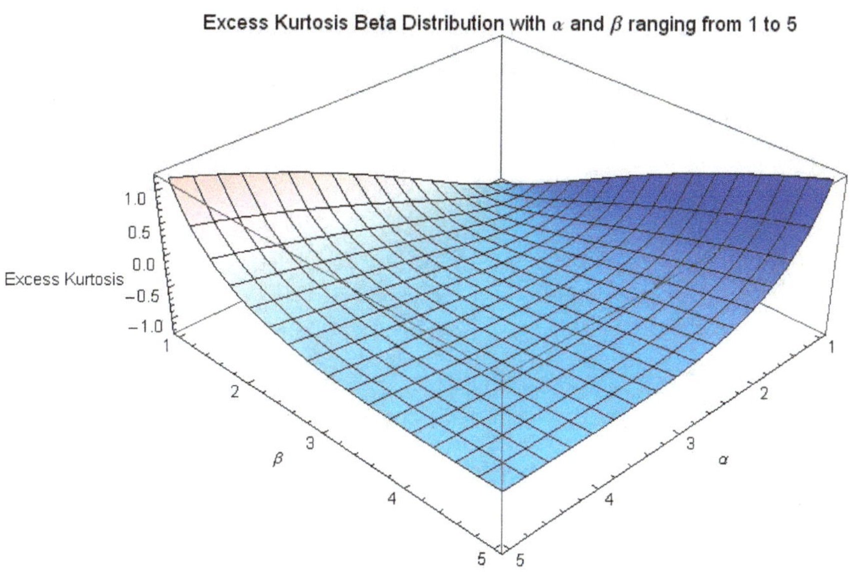

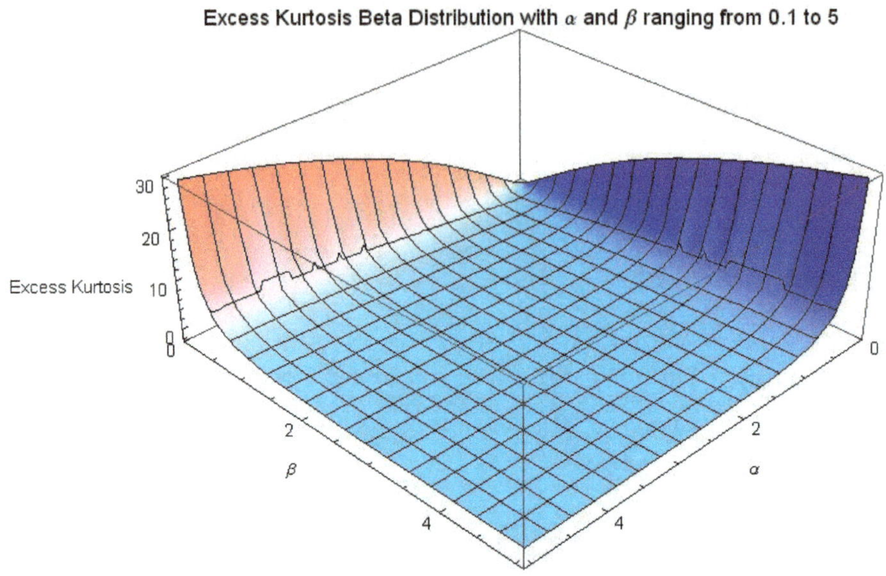

Characteristic Function

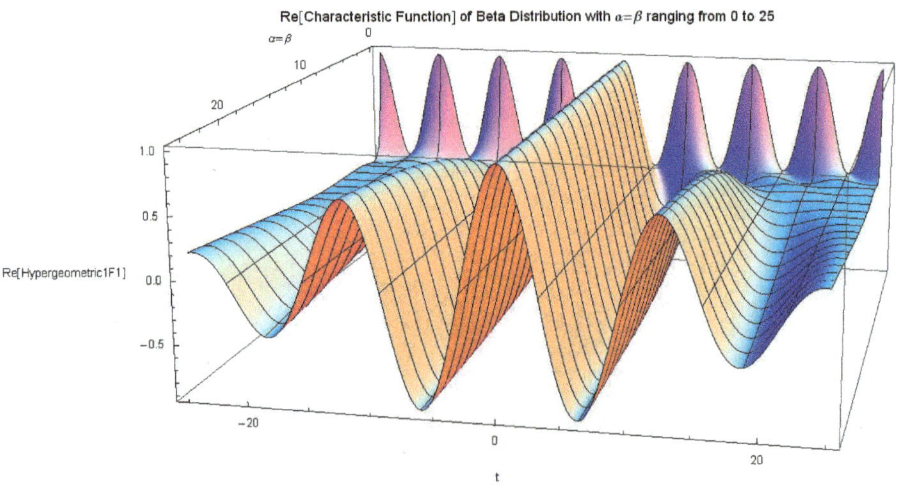

Re(characteristic function) symmetric case $\alpha = \beta$ ranging from 25 to 0

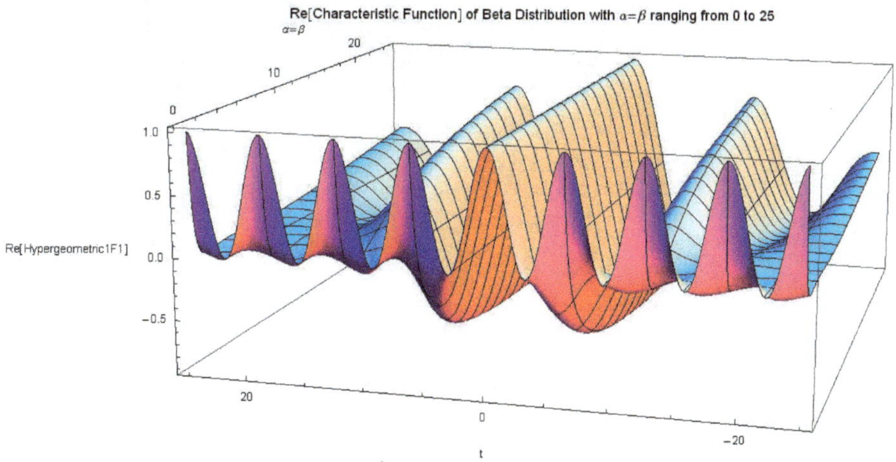

Re(characteristic function) symmetric case α = β ranging from 0 to 25.

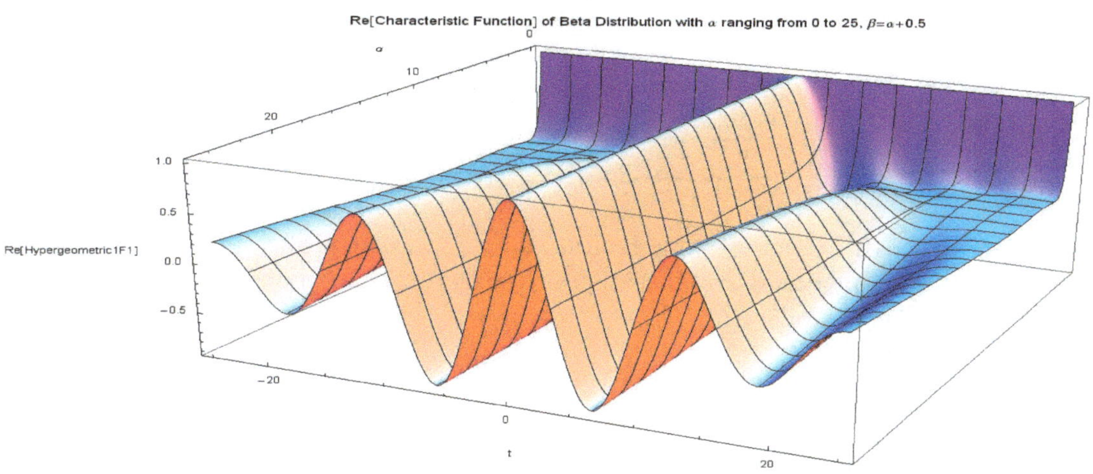

Re(characteristic function) β = α + 1/2; α ranging from 25 to 0.

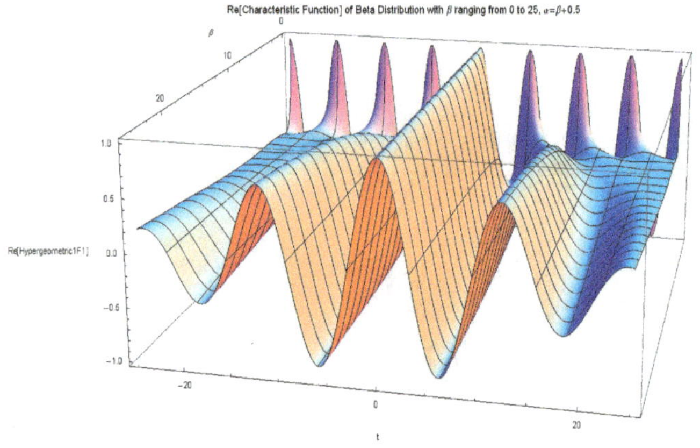

Re(characteristic function) α = β + 1/2; β ranging from 25 to 0

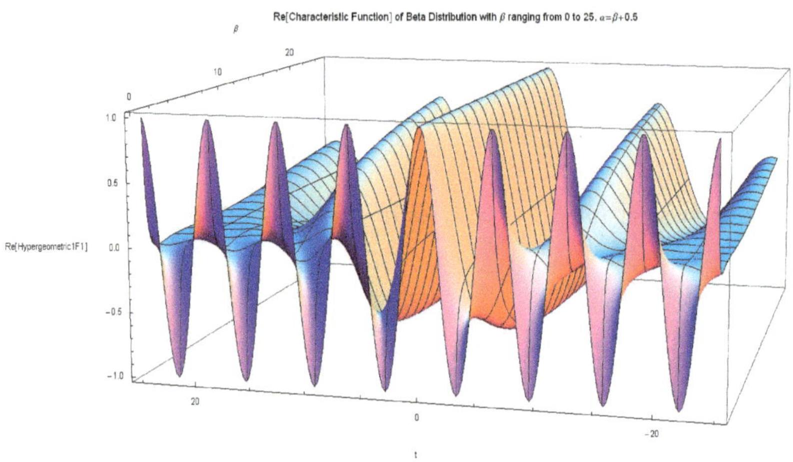

Re(characteristic function) α = β + 1/2; β ranging from 0 to 25.

The characteristic function is the Fourier transform of the probability density function. The characteristic function of the beta distribution is Kummer's confluent hypergeometric function (of the first kind):

$$\begin{aligned}
\varphi_X(\alpha;\beta;t) &= \mathbf{E}\left[e^{itX}\right] \\
&= \int_0^1 e^{itx} f(x;\alpha,\beta)dx \\
&= {}_1F_1(\alpha;\alpha+\beta;it) \\
&= \sum_{n=0}^{\infty} \frac{\alpha^{(n)}(it)^n}{(\alpha+\beta)^{(n)}n!} \\
&= 1+\sum_{k=1}^{\infty}\left(\prod_{r=0}^{k-1}\frac{\alpha+r}{\alpha+\beta+r}\right)\frac{(it)^k}{k!}
\end{aligned}$$

where

$x^{(n)} = x(x+1)(x+2)\cdots(x+n-1)$

is the rising factorial, also called the "Pochhammer symbol". The value of the characteristic function for t = 0, is one:

$$\varphi_X(\alpha;\beta;0) = {}_1F_1(\alpha;\alpha+\beta;0) = 1.$$

Also, the real and imaginary parts of the characteristic function enjoy the following symmetries with respect to the origin of variable t:

$$\mathrm{Re}\left[{}_1F_1(\alpha;\alpha+\beta;it)\right] = \mathrm{Re}\left[{}_1F_1(\alpha;\alpha+\beta;-it)\right]$$
$$\mathrm{Im}\left[{}_1F_1(\alpha;\alpha+\beta;it)\right] = -\mathrm{Im}\left[{}_1F_1(\alpha;\alpha+\beta;-it)\right]$$

The symmetric case α = β simplifies the characteristic function of the beta distribution to a Bessel function, since in the special case α + β = 2α the confluent hypergeometric function (of the first kind) reduces to a Bessel function (the modified Bessel function of the first kind $I_{\alpha-\frac{1}{2}}$ using Kummer's second transformation as follows:

$$\begin{aligned}{}_1F_1(\alpha;2\alpha;it) &= e^{\frac{it}{2}}\,{}_0F_1\left(;\alpha+\tfrac{1}{2};\frac{(it)^2}{16}\right)\\ &= e^{\frac{it}{2}}\left(\frac{it}{4}\right)^{\frac{1}{2}-\alpha}\Gamma\left(\alpha+\tfrac{1}{2}\right)I_{\alpha-\frac{1}{2}}\left(\frac{it}{2}\right)\end{aligned}$$

In the accompanying plots, the real part (Re) of the characteristic function of the beta distribution is displayed for symmetric (α = β) and skewed (α ≠ β) cases.

Other Moments

Moment Generating Function

It also follows that the moment generating function is

$$\begin{aligned}M_X(\alpha;\beta;t) &= \mathrm{E}\left[e^{tX}\right]\\ &= \int_0^1 e^{tx} f(x;\alpha,\beta)\,dx\\ &= {}_1F_1(\alpha;\alpha+\beta;t)\\ &= \sum_{n=0}^{\infty}\frac{\alpha^{(n)}}{(\alpha+\beta)^{(n)}}\frac{t^n}{n!}\\ &= 1+\sum_{k=1}^{\infty}\left(\prod_{r=0}^{k-1}\frac{\alpha+r}{\alpha+\beta+r}\right)\frac{t^k}{k!}\end{aligned}$$

In particular $M_X(\alpha;\beta;0) = 1$.

Higher Moments

Using the moment generating function, the k-th raw moment is given by the factor

$$\prod_{r=0}^{k-1} \frac{\alpha+r}{\alpha+\beta+r}$$

multiplying the (exponential series) term $\left(\frac{t^k}{k!}\right)$ in the series of the moment generating function

$$\mathrm{E}[X^k] = \frac{\alpha^{(k)}}{(\alpha+\beta)^{(k)}} = \prod_{r=0}^{k-1} \frac{\alpha+r}{\alpha+\beta+r}$$

where $(x)^{(k)}$ is a Pochhammer symbol representing rising factorial. It can also be written in a recursive form as

$$\mathrm{E}[X^k] = \frac{\alpha+k-1}{\alpha+\beta+k-1}\mathrm{E}[X^{k-1}].$$

Since the moment generating function $M_X(\alpha;\beta;\cdot)$ has a positive radius of convergence, the beta distribution is determined by its moments.

Moments of Transformed Random Variables

Moments of Linearly Transformed, Product and Inverted Random Variables

One can also show the following expectations for a transformed random variable, where the random variable X is Beta-distributed with parameters α and β: X ~ Beta(α, β). The expected value of the variable (1−X) is the mirror-symmetry of the expected value based on X:

$$\mathrm{E}[1-X] = \frac{\beta}{\alpha+\beta}$$

$$\mathrm{E}[X(1-X)] = \mathrm{E}[(1-X)X] = \frac{\alpha\beta}{(\alpha+\beta)(\alpha+\beta+1)}$$

Due to the mirror-symmetry of the probability density function of the beta distribution, the variances based on variables X and (1−X) are identical, and the covariance on X(1-X) is the negative of the variance:

$$\mathrm{var}[(1-X)] = \mathrm{var}[X] = -\mathrm{cov}[X,(1-X)] = \frac{\alpha\beta}{(\alpha+\beta)^2(\alpha+\beta+1)}$$

These are the expected values for inverted variables, (these are related to the harmonic means, see section titled "Harmonic mean"):

$$E\left[\frac{1}{X}\right] = \frac{\alpha+\beta-1}{\alpha-1} \text{ if } \alpha > 1$$
$$E\left[\frac{1}{1-X}\right] = \frac{\alpha+\beta-1}{\beta-1} \text{ if } \beta > 1$$

The following transformation by dividing the variable X by its mirror-image X/(1−X) results in the expected value of the "inverted beta distribution" or beta prime distribution (also known as beta distribution of the second kind or Pearson›s Type VI):

$$E\left[\frac{X}{1-X}\right] = \frac{\alpha}{\beta-1} \text{ if } \beta > 1$$
$$E\left[\frac{1-X}{X}\right] = \frac{\beta}{\alpha-1} \text{ if } \alpha > 1$$

Variances of these transformed variables can be obtained by integration, as the expected values of the second moments centered on the corresponding variables:

$$\operatorname{var}\left[\frac{1}{X}\right] = E\left[\left(\frac{1}{X} - E\left[\frac{1}{X}\right]\right)^2\right] =$$
$$\operatorname{var}\left[\frac{1-X}{X}\right] = E\left[\left(\frac{1-X}{X} - E\left[\frac{1-X}{X}\right]\right)^2\right] = \frac{\beta(\alpha+\beta-1)}{(\alpha-2)(\alpha-1)^2} \text{ if } \alpha > 2$$

The following variance of the variable X divided by its mirror-image (X/(1−X) results in the variance of the "inverted beta distribution" or beta prime distribution (also known as beta distribution of the second kind or Pearson›s Type VI):

$$\operatorname{var}\left[\frac{1}{1-X}\right] = E\left[\left(\frac{1}{1-X} - E\left[\frac{1}{1-X}\right]\right)^2\right] = \operatorname{var}\left[\frac{X}{1-X}\right] =$$
$$E\left[\left(\frac{X}{1-X} - E\left[\frac{X}{1-X}\right]\right)^2\right] = \frac{\alpha(\alpha+\beta-1)}{(\beta-2)(\beta-1)^2} \text{ if } \beta > 2$$

The covariances are:

$$\operatorname{cov}\left[\frac{1}{X}, \frac{1}{1-X}\right] = \operatorname{cov}\left[\frac{1-X}{X}, \frac{X}{1-X}\right] = \operatorname{cov}\left[\frac{1}{X}, \frac{X}{1-X}\right] = \operatorname{cov}\left[\frac{1-X}{X}, \frac{1}{1-X}\right] = \frac{\alpha+\beta-1}{(\alpha-1)(\beta-1)} \text{ if } \alpha, \beta > 1$$

These expectations and variances appear in the four-parameter Fisher information matrix (section titled "Fisher information," "four parameters").

Moments of Logarithmically Transformed Random Variables

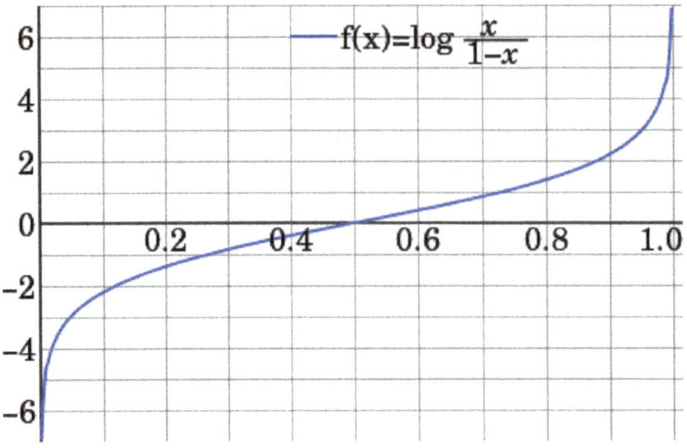

Plot of logit(X) = ln(X/(1−X)) (vertical axis) vs. X in the domain of 0 to 1 (horizontal axis). Logit transformations are interesting, as they usually transform various shapes (including J-shapes) into (usually skewed) bell-shaped densities over the logit variable, and they may remove the end singularities over the original variable.

Expected values for logarithmic transformations (useful for maximum likelihood estimates, see section titled "Parameter estimation, Maximum likelihood" below) are discussed in this section. The following logarithmic linear transformations are related to the geometric means G_X and $G_{(1-X)}$ (see section titled "Geometric mean"):

$$E[\ln(X)] = \psi(\alpha) - \psi(\alpha+\beta) = -E\left[\ln\left(\frac{1}{X}\right)\right],$$

$$E[\ln(1-X)] = \psi(\beta) - \psi(\alpha+\beta) = -E\left[\ln\left(\frac{1}{1-X}\right)\right].$$

Where the **digamma function** $\psi(\alpha)$ is defined as the logarithmic derivative of the gamma function:

$$\psi(\alpha) = \frac{d\ln\Gamma(\alpha)}{d\alpha}$$

Logit transformations are interesting, as they usually transform various shapes (including J-shapes) into (usually skewed) bell-shaped densities over the logit variable, and they may remove the end singularities over the original variable:

$$E\left[\ln\left(\frac{X}{1-X}\right)\right] = \psi(\alpha) - \psi(\beta) = E[\ln(X)] + E\left[\ln\left(\frac{1}{1-X}\right)\right],$$

$$E\left[\ln\left(\frac{1-X}{X}\right)\right] = \psi(\beta) - \psi(\alpha) = -E\left[\ln\left(\frac{X}{1-X}\right)\right].$$

Johnson considered the distribution of the logit - transformed variable ln(X/1−X), including its moment generating function and approximations for large values of the shape parameters. This transformation extends the finite support [0, 1] based on the original variable X to infinite support in both directions of the real line (−∞, +∞).

Higher order logarithmic moments can be derived by using the representation of a beta distribution as a proportion of two Gamma distributions and differentiating through the integral. They can be expressed in terms of higher order poly-gamma functions as follows:

$$\begin{aligned}
\mathrm{E}\left[\ln^2(X)\right] &= (\psi(\alpha)-\psi(\alpha+\beta))^2+\psi_1(\alpha)-\psi_1(\alpha+\beta),\\
\mathrm{E}\left[\ln^2(1-X)\right] &= (\psi(\beta)-\psi(\alpha+\beta))^2+\psi_1(\beta)-\psi_1(\alpha+\beta),\\
\mathrm{E}[\ln(X)\ln(1-X)] &= (\psi(\alpha)-\psi(\alpha+\beta))(\psi(\beta)-\psi(\alpha+\beta))-\psi_1(\alpha+\beta).
\end{aligned}$$

therefore the variance of the logarithmic variables and covariance of ln(X) and ln(1−X) are:

$$\operatorname{cov}[\ln(X),\ln(1-X)] = \mathrm{E}[\ln(X)\ln(1-X)] - \mathrm{E}[\ln(X)]\,\mathrm{E}[\ln(1-X)] = -\psi_1(\alpha+\beta)$$

$$\begin{aligned}
\operatorname{var}[\ln X] &= \mathrm{E}[\ln^2(X)] - (\mathrm{E}[\ln(X)])^2\\
&= \psi_1(\alpha)-\psi_1(\alpha+\beta)\\
&= \psi_1(\alpha)+\operatorname{cov}[\ln(X),\ln(1-X)]
\end{aligned}$$

$$\begin{aligned}
\operatorname{var}[\ln(1-X)] &= \mathrm{E}[\ln^2(1-X)] - (\mathrm{E}[\ln(1-X)])^2\\
&= \psi_1(\beta)-\psi_1(\alpha+\beta)\\
&= \psi_1(\beta)+\operatorname{cov}[\ln(X),\ln(1-X)]
\end{aligned}$$

where the **trigamma function**, denoted $\psi_1(\alpha)$, is the second of the polygamma functions, and is defined as the derivative of the digamma function:

$$\psi_1(\alpha) = \frac{d^2\ln\Gamma(\alpha)}{d\alpha^2} = \frac{d\psi(\alpha)}{d\alpha}.$$

The variances and covariance of the logarithmically transformed variables X and (1−X) are different, in general, because the logarithmic transformation destroys the mirror-symmetry of the original variables X and (1−X), as the logarithm approaches negative infinity for the variable approaching zero.

These logarithmic variances and covariance are the elements of the Fisher information matrix for the beta distribution. They are also a measure of the curvature of the log likelihood function (see section on Maximum likelihood estimation).

The variances of the log inverse variables are identical to the variances of the log variables:

$$\begin{aligned}
\operatorname{var}\left[\ln\left(\frac{1}{X}\right)\right] &= \operatorname{var}[\ln(X)] = \psi_1(\alpha)-\psi_1(\alpha+\beta),\\
\operatorname{var}\left[\ln\left(\frac{1}{1-X}\right)\right] &= \operatorname{var}[\ln(1-X)] = \psi_1(\beta)-\psi_1(\alpha+\beta),\\
\operatorname{cov}\left[\ln\left(\frac{1}{X}\right),\ln\left(\frac{1}{1-X}\right)\right] &= \operatorname{cov}[\ln(X),\ln(1-X)] = -\psi_1(\alpha+\beta).
\end{aligned}$$

It also follows that the variances of the logit transformed variables are:

$$\operatorname{var}\left[\ln\left(\frac{X}{1-X}\right)\right] = \operatorname{var}\left[\ln\left(\frac{1-X}{X}\right)\right] = -\operatorname{cov}\left[\ln\left(\frac{X}{1-X}\right),\ln\left(\frac{1-X}{X}\right)\right] = \psi_1(\alpha)+\psi_1(\beta)$$

Quantities of Informaation (Entropy)

Given a beta distributed random variable, X ~ Beta(α, β), the differential entropy of X is(measured in nats), the expected value of the negative of the logarithm of the probability density function:

$$\begin{aligned} h(X) &= \mathrm{E}[-\ln(f(x;\alpha,\beta))] \\ &= \int_0^1 -f(x;\alpha,\beta)\ln(f(x;\alpha,\beta))dx \\ &= \ln(\mathrm{B}(\alpha,\beta)) - (\alpha-1)\psi(\alpha) - (\beta-1)\psi(\beta) + (\alpha+\beta-2)\psi(\alpha+\beta) \end{aligned}$$

where f(x; α, β) is the probability density function of the beta distribution:

$$f(x;\alpha,\beta) = \frac{1}{\mathrm{B}(\alpha,\beta)}x^{\alpha-1}(1-x)^{\beta-1}$$

The digamma function ψ appears in the formula for the differential entropy as a consequence of Euler's integral formula for the harmonic numbers which follows from the integral:

$$\int_0^1 \frac{1-x^{\alpha-1}}{1-x}dx = \psi(\alpha) - \psi(1)$$

The differential entropy of the beta distribution is negative for all values of α and β greater than zero, except at α = β = 1 (for which values the beta distribution is the same as the uniform distribution), where the differential entropy reaches its maximum value of zero. It is to be expected that the maximum entropy should take place when the beta distribution becomes equal to the uniform distribution, since uncertainty is maximal when all possible events are equiprobable.

For α or β approaching zero, the differential entropy approaches its minimum value of negative infinity. For (either or both) α or β approaching zero, there is a maximum amount of order: all the probability density is concentrated at the ends, and there is zero probability density at points located between the ends. Similarly for (either or both) α or β approaching infinity, the differential entropy approaches its minimum value of negative infinity, and a maximum amount of order. If either α or β approaches infinity (and the other is finite) all the probability density is concentrated at an end, and the probability density is zero everywhere else. If both shape parameters are equal (the symmetric case), α = β, and they approach infinity simultaneously, the probability density becomes a spike (Dirac delta function) concentrated at the middle x = 1/2, and hence there is 100% probability at the middle x = 1/2 and zero probability everywhere else.

Diff. Entropy Beta Distribution with α and β ranging from 1 to 5

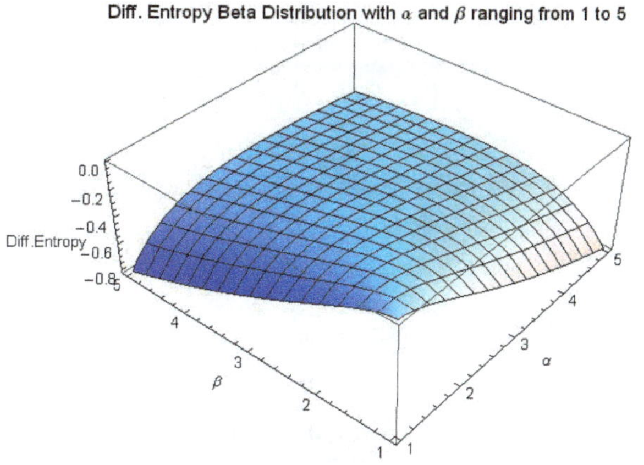

Diff. Entropy Beta Distribution with α and β ranging from 0.1 to 5

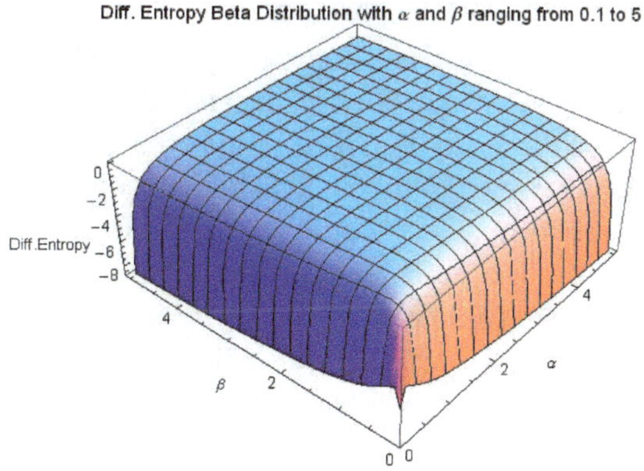

The (continuous case) differential entropy was introduced by Shannon in his original paper (where he named it the "entropy of a continuous distribution"), as the concluding part of the same paper where he defined the discrete entropy. It is known since then that the differential entropy may differ from the infinitesimal limit of the discrete entropy by an infinite offset, therefore the differential entropy can be negative (as it is for the beta distribution). What really matters is the relative value of entropy.

Given two beta distributed random variables, $X_1 \sim \mathrm{Beta}(\alpha, \beta)$ and $X_2 \sim \mathrm{Beta}(\alpha', \beta')$, the cross entropy is (measured in nats)

$$\begin{aligned} H(X_1, X_2) &= \int_0^1 -f(x;\alpha,\beta)\ln(f(x;\alpha',\beta'))dx \\ &= \ln(\mathrm{B}(\alpha',\beta')) - (\alpha'-1)\psi(\alpha) - (\beta'-1)\psi(\beta) + (\alpha'+\beta'-2)\psi(\alpha+\beta). \end{aligned}$$

The cross entropy has been used as an error metric to measure the distance between two hypotheses. Its absolute value is minimum when the two distributions are identical. It is the information measure most closely related to the log maximum likelihood (see section on "Parameter estimation. Maximum likelihood estimation")).

The relative entropy, or Kullback–Leibler divergence $D_{KL}(X_1, X_2)$, is a measure of the inefficiency of assuming that the distribution is $X_2 \sim \mathrm{Beta}(\alpha', \beta')$ when the distribution is really $X_1 \sim \mathrm{Beta}(\alpha, \beta)$. It is defined as follows (measured in nats).

$$\begin{aligned}
D_{\mathrm{KL}}(X_1, X_2) &= \int_0^1 f(x;\alpha,\beta)\ln\left(\frac{f(x;\alpha,\beta)}{f(x;\alpha',\beta')}\right)dx \\
&= \left(\int_0^1 f(x;\alpha,\beta)\ln(f(x;\alpha,\beta))dx\right) - \left(\int_0^1 f(x;\alpha,\beta)\ln(f(x;\alpha',\beta'))dx\right) \\
&= -h(X_1) + H(X_1, X_2) \\
&= \ln\left(\frac{\mathrm{B}(\alpha',\beta')}{\mathrm{B}(\alpha,\beta)}\right) + (\alpha-\alpha')\psi(\alpha) + (\beta-\beta')\psi(\beta) + (\alpha'-\alpha+\beta'-\beta)\psi(\alpha+\beta).
\end{aligned}$$

The relative entropy, or Kullback–Leibler divergence, is always non-negative. A few numerical examples follow:

- $X_1 \sim$ Beta(1, 1) and $X_2 \sim$ Beta(3, 3); $D_{KL}(X_1, X_2) = 0.598803$; $D_{KL}(X_2, X_1) = 0.267864$; $h(X_1) = 0$; $h(X_2) = -0.267864$
- $X_1 \sim$ Beta(3, 0.5) and $X_1 \sim$ Beta(0.5, 3); $D_{KL}(X_1, X_2) = 7.21574$; $D_{KL}(X_2, X_1) = 7.21574$; $h(X_1) = -1.10805$; $h(X_2) = -1.10805$.

The Kullback–Leibler divergence is not symmetric $D_{KL}(X_1, X_2) \neq D_{KL}(X_2, X_1)$ for the case in which the individual beta distributions Beta(1, 1) and Beta(3, 3) are symmetric, but have different entropies $h(X_1) \neq h(X_2)$. The value of the Kullback divergence depends on the direction traveled: whether going from a higher (differential) entropy to a lower (differential) entropy or the other way around. In the numerical example above, the Kullback divergence measures the inefficiency of assuming that the distribution is (bell-shaped) Beta(3, 3), rather than (uniform) Beta(1, 1). The "h" entropy of Beta(1, 1) is higher than the "h" entropy of Beta(3, 3) because the uniform distribution Beta(1, 1) has a maximum amount of disorder. The Kullback divergence is more than two times higher (0.598803 instead of 0.267864) when measured in the direction of decreasing entropy: the direction that assumes that the (uniform) Beta(1, 1) distribution is (bell-shaped) Beta(3, 3) rather than the other way around. In this restricted sense, the Kullback divergence is consistent with the second law of thermodynamics.

The Kullback–Leibler divergence is symmetric $D_{KL}(X_1, X_2) = D_{KL}(X_2, X_1)$ for the skewed cases Beta(3, 0.5) and Beta(0.5, 3) that have equal differential entropy $h(X_1) = h(X_2)$.

The symmetry condition:

$$D_{\mathrm{KL}}(X_1, X_2) = D_{\mathrm{KL}}(X_2, X_1), \text{ if } h(X_1) = h(X_2), \text{ for (skewed) } \alpha \neq \beta$$

follows from the above definitions and the mirror-symmetry $f(x; \alpha, \beta) = f(1-x; \alpha, \beta)$ enjoyed by the beta distribution.

Relationships between Statistical Measures

Mean, mode and Median Relationship

If $1 < \alpha < \beta$ then mode $\leq$ median $\leq$ mean. Expressing the mode (only for $\alpha, \beta > 1$), and the mean in terms of α and β:

$$\frac{\alpha-1}{\alpha+\beta-2} \leq \text{median} \leq \frac{\alpha}{\alpha+\beta},$$

If $1 < \beta < \alpha$ then the order of the inequalities are reversed. For $\alpha, \beta > 1$ the absolute distance between the mean and the median is less than 5% of the distance between the maximum and minimum values of x. On the other hand, the absolute distance between the mean and the mode can reach 50% of the distance between the maximum and minimum values of x, for the (pathological) case of $\alpha = 1$ and $\beta = 1$ (for which values the beta distribution approaches the uniform distribution and the differential entropy approaches its maximum value, and hence maximum "disorder").

For example, for $\alpha = 1.0001$ and $\beta = 1.00000001$:

- mode = 0.9999; PDF(mode) = 1.00010
- mean = 0.500025; PDF(mean) = 1.00003
- median = 0.500035; PDF(median) = 1.00003
- mean − mode = −0.499875
- mean − median = -9.65538×10^{-6}

(where PDF stands for the value of the probability density function)

(Mean – Median) Beta Distribution with α and β ranging from 1 to 5

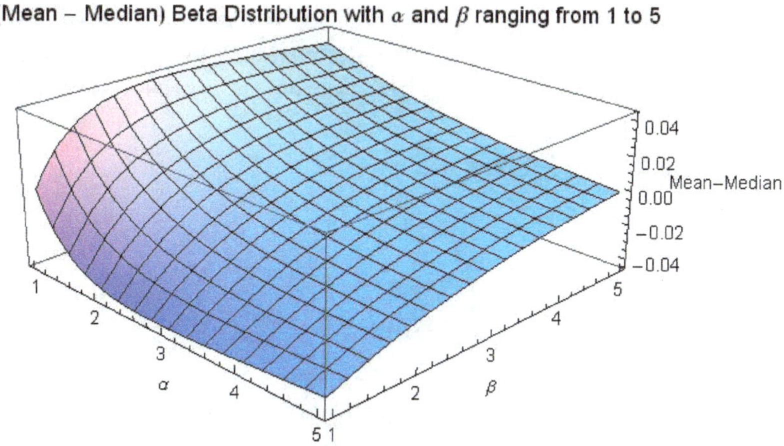

(Mean – Mode) Beta Distribution with α and β ranging from 1 to 5

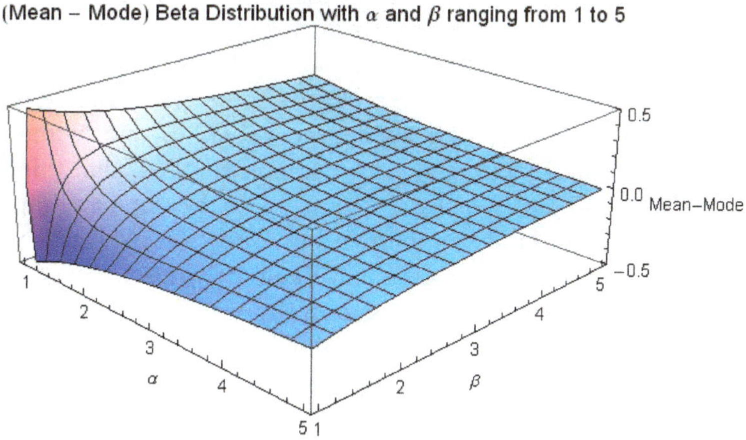

Mean, Geometric mean and Harmonic Mean Relationship

:Mean, Median, Geometric Mean and Harmonic Mean for Beta distribution with $0 < \alpha = \beta < 5$

It is known from the inequality of arithmetic and geometric means that the geometric mean is lower than the mean. Similarly, the harmonic mean is lower than the geometric mean. The accompanying plot shows that for $\alpha = \beta$, both the mean and the median are exactly equal to 1/2, regardless of the value of $\alpha = \beta$, and the mode is also equal to 1/2 for $\alpha = \beta > 1$, however the geometric and harmonic means are lower than 1/2 and they only approach this value asymptotically as $\alpha = \beta \to \infty$.

Kurtosis Bounded by the Square of the Skewness

Beta distribution α and β parameters vs. excess Kurtosis and squared Skewness

As remarked by Feller, in the Pearson system the beta probability density appears as type I (any difference between the beta distribution and Pearson's type I distribution is only superficial and it makes no difference for the following discussion regarding the relationship between kurtosis and skewness). Karl Pearson showed, in Plate 1 of his paper published in 1916, a graph with the kurtosis as the vertical axis (ordinate) and the square of the skewness as the horizontal axis (abscissa), in which a number of distributions were displayed. The region occupied by the beta distribution is bounded by the following two lines in the (skewness^2,kurtosis) plane, or the (skewness^2,excess kurtosis) plane:

$$(\text{skewness})^2 + 1 < \text{kurtosis} < \frac{3}{2}(\text{skewness})^2 + 3$$

or, equivalently,

$$(\text{skewness})^2 - 2 < \text{excess kurtosis} < \frac{3}{2}(\text{skewness})^2$$

(At a time when there were no powerful digital computers), Karl Pearson accurately computed further boundaries, for example, separating the "U-shaped" from the "J-shaped" distributions. The lower boundary line (excess kurtosis + 2 − skewness^2 = 0) is produced by skewed "U-shaped" beta distributions with both values of shape parameters α and β close to zero. The upper boundary line (excess kurtosis − (3/2) skewness^2 = 0) is produced by extremely skewed distributions with very large values of one of the parameters and very small values of the other parameter. Karl Pearson showed that this upper boundary line (excess kurtosis − (3/2) skewness^2 = 0) is also the intersection with Pearson's distribution III, which has unlimited support in one direction (towards positive infinity), and can be bell-shaped or J-shaped. His son, Egon Pearson, showed that the region (in the kurtosis/squared-skewness plane) occupied by the beta distribution (equivalently, Pearson›s distribution I) as it approaches this boundary (excess kurtosis − (3/2) skewness^2 = 0) is shared with the noncentral chi-squared distribution. Karl Pearson (Pearson 1895, pp. 357, 360, 373–376) also showed that the gamma distribution is a Pearson type III distribution. Hence this boundary line for Pearson›s type III distribution is known as the gamma line. (This can be shown from the fact that the excess kurtosis of the gamma distribution is 6/k and the square of the skewness is 4/k, hence (excess kurtosis − (3/2) skewness^2 = 0) is identically satisfied by the gamma distribution regardless of the value of the parameter "k"). Pearson later noted that the chi-squared distribution is a special case of Pearson's type III and also shares this boundary line (as it is apparent from the fact that for the chi-squared distribution the excess kurtosis is 12/k and the square of the skewness is 8/k, hence (excess kurtosis − (3/2) skewness^2 = 0) is identically satisfied regardless of the value of the parameter "k"). This is to be expected, since the chi-squared distribution $X \sim \chi^2(k)$ is a special case of the gamma distribution, with parametrization $X \sim \Gamma(k/2, 1/2)$ where k is a positive integer that specifies the "number of degrees of freedom" of the chi-squared distribution.

An example of a beta distribution near the upper boundary (excess kurtosis − (3/2) skewness^2 = 0) is given by $\alpha = 0.1$, $\beta = 1000$, for which the ratio (excess kurtosis)/(skewness^2) = 1.49835 approaches the upper limit of 1.5 from below. An example of a beta distribution near the lower boundary (excess kurtosis + 2 − skewness^2 = 0) is given by $\alpha = 0.0001$, $\beta = 0.1$, for which values the expression (excess kurtosis + 2)/(skewness^2) = 1.01621 approaches the lower limit of 1 from above. In the infinitesimal limit for both α and β approaching zero symmetrically, the excess kurtosis reaches its minimum value at −2. This minimum value occurs at the point at which the lower boundary line intersects the vertical axis (ordinate). (However, in Pearson's original chart, the ordinate is kurtosis, instead of excess kurtosis, and it increases downwards rather than upwards).

Values for the skewness and excess kurtosis below the lower boundary (excess kurtosis + 2 − skewness^2 = 0) cannot occur for any distribution, and hence Karl Pearson appropriately called the region below this boundary the "impossible region." The boundary for this "impossible region" is determined by (symmetric or skewed) bimodal "U"-shaped distributions for which parameters α and β approach zero and hence all the probability density is concentrated at the ends: x = 0, 1 with practically nothing in between them. Since for $\alpha \approx \beta \approx 0$ the probability density is concentrated at the two ends x = 0 and x = 1, this "impossible boundary" is determined by a 2-point distribution: the probability can only take 2 values (Bernoulli distribution), one value with probability p and the other with probability q = 1−p. For cases approaching this limit boundary with symmetry α = β, skewness ≈ 0, excess kurtosis ≈ −2 (this is the lowest excess kurtosis possible for any distribution), and the probabilities are $p \approx q \approx 1/2$. For cases approaching this limit boundary with skewness, excess kurtosis ≈ −2 + skewness^2, and the probability density is concentrated more at one end than the other end (with practically nothing in between), with probabilities $p = \frac{\beta}{\alpha+\beta}$ at the left end x = 0 and $q = 1 - p = \frac{\alpha}{\alpha+\beta}$ at the right end x = 1.

Symmetry

All statements are conditional on α, β > 0

- Probability density function reflection symmetry

$$f(x;\alpha,\beta) = f(1-x;\beta,\alpha)$$

- Cumulative distribution function reflection symmetry plus unitary translation

$$F(x;\alpha,\beta) = I_x(\alpha,\beta) = 1 - F(1-x;\beta,\alpha) = 1 - I_{1-x}(\beta,\alpha)$$

- Mode reflection symmetry plus unitary translation

$$\text{mode}(\mathrm{B}(\alpha,\beta)) = 1 - \text{mode}(\mathrm{B}(\beta,\alpha)),\ \text{if}\ \mathrm{B}(\beta,\alpha) \neq \mathrm{B}(1,1)$$

- Median reflection symmetry plus unitary translation

$$\text{median}(\mathrm{B}(\alpha,\beta)) = 1 - \text{median}(\mathrm{B}(\beta,\alpha))$$

- Mean reflection symmetry plus unitary translation

$$\mu(\mathrm{B}(\alpha,\beta)) = 1 - \mu(\mathrm{B}(\beta,\alpha))$$

- Harmonic means each is individually asymmetric, the following symmetry applies between the harmonic mean based on X and the harmonic mean based on its reflection (1-X)

$$H_X(\mathrm{B}(\alpha,\beta)) = H_{(1-X)}(\mathrm{B}(\beta,\alpha))\ \text{if}\ \alpha,\beta > 1\,.$$

- Variance symmetry

$$\text{var}(\mathrm{B}(\alpha,\beta)) = \text{var}(\mathrm{B}(\beta,\alpha))$$

- Geometric variances each is individually asymmetric, the following symmetry applies between the log geometric variance based on X and the log geometric variance based on its reflection (1-X)

$$\ln(\text{var}_{\mathrm{GX}}(\mathrm{B}(\alpha,\beta))) = \ln(\text{var}_{\mathrm{G(1\text{-}X)}}(\mathrm{B}(\beta,\alpha)))$$

- Geometric covariance symmetry

$$\ln \text{cov}_{\mathrm{GX,(1\text{-}X)}}(\mathrm{B}(\alpha,\beta)) = \ln \text{cov}_{\mathrm{GX,(1\text{-}X)}}(\mathrm{B}(\beta,\alpha))$$

- Mean absolute deviation around the mean symmetry

$$\mathrm{E}[|X - E[X]|](\mathrm{B}(\alpha,\beta)) = \mathrm{E}[|X - E[X]|](\mathrm{B}(\beta,\alpha))$$

- Skewness skew-symmetry

$$\text{skewness}(\mathrm{B}(\alpha,\beta)) = -\,\text{skewness}(\mathrm{B}(\beta,\alpha))$$

- Excess kurtosis symmetry

$$\text{excess kurtosis}(\mathrm{B}(\alpha,\beta)) = \text{excess kurtosis}(\mathrm{B}(\beta,\alpha))$$

- Characteristic function symmetry of Real part (with respect to the origin of variable "t")

$$\mathrm{Re}[{}_1F_1(\alpha;\alpha+\beta;it)] = \mathrm{Re}[{}_1F_1(\alpha;\alpha+\beta;-it)]$$

- Characteristic function skew-symmetry of Imaginary part (with respect to the origin of variable "t")

$$\mathrm{Im}[{}_1F_1(\alpha;\alpha+\beta;it)] = -\mathrm{Im}[{}_1F_1(\alpha;\alpha+\beta;-it)]$$

- Characteristic function symmetry of Absolute value (with respect to the origin of variable "t")

$$\mathrm{Abs}[{}_1F_1(\alpha;\alpha+\beta;it)] = \mathrm{Abs}[{}_1F_1(\alpha;\alpha+\beta;-it)]$$

- Differential entropy symmetry

$$h(\mathrm{B}(\alpha,\beta)) = h(\mathrm{B}(\beta,\alpha))$$

- Relative Entropy (also called Kullback–Leibler divergence) symmetry

$$D_{\mathrm{KL}}(X_1,X_2) = D_{\mathrm{KL}}(X_2,X_1), \text{ if } h(X_1) = h(X_2), \text{ for (skewed) } \alpha \neq \beta$$

- Fisher information matrix symmetry

$$\mathcal{I}_{i,j} = \mathcal{I}_{j,i}$$

Geometry of the Probability Density Function

Inflection Points

Inflection point location versus α and β showing regions with one inflection point

Inflection point location versus α and β showing region with two inflection points

For certain values of the shape parameters α and β, the probability density function has inflection points, at which the curvature changes sign. The position of these inflection points can be useful as a measure of the dispersion or spread of the distribution.

Defining the following quantity:

$$\kappa = \frac{\sqrt{\frac{(\alpha-1)(\beta-1)}{\alpha+\beta-3}}}{\alpha+\beta-2}$$

Points of inflection occur, depending on the value of the shape parameters α and β, as follows:

- (α > 2, β > 2) The distribution is bell-shaped (symmetric for α = β and skewed otherwise), with **two inflection points**, equidistant from the mode:

$$x = \text{mode} \pm \kappa = \frac{\alpha - 1 \pm \sqrt{\frac{(\alpha-1)(\beta-1)}{\alpha+\beta-3}}}{\alpha+\beta-2}$$

- (α = 2, β > 2) The distribution is unimodal, positively skewed, right-tailed, with **one inflection point**, located to the right of the mode:

$$x = \text{mode} + \kappa = \frac{2}{\beta}$$

- ($\alpha > 2$, $\beta = 2$) The distribution is unimodal, negatively skewed, left-tailed, with **one inflection point**, located to the left of the mode:

$$x = \text{mode} - \kappa = 1 - \frac{2}{\alpha}$$

- ($1 < \alpha < 2$, $\beta > 2$, $\alpha+\beta>2$) The distribution is unimodal, positively skewed, right-tailed, with **one inflection point**, located to the right of the mode:

$$x = \text{mode} + \kappa = \frac{\alpha - 1 + \sqrt{\frac{(\alpha-1)(\beta-1)}{\alpha+\beta-3}}}{\alpha + \beta - 2}$$

- ($0 < \alpha < 1$, $1 < \beta < 2$) The distribution has a mode at the left end $x = 0$ and it is positively skewed, right-tailed. There is **one inflection point**, located to the right of the mode:

$$x = \frac{\alpha - 1 + \sqrt{\frac{(\alpha-1)(\beta-1)}{\alpha+\beta-3}}}{\alpha + \beta - 2}$$

- ($\alpha > 2$, $1 < \beta < 2$) The distribution is unimodal negatively skewed, left-tailed, with **one inflection point**, located to the left of the mode:

$$x = \text{mode} - \kappa = \frac{\alpha - 1 - \sqrt{\frac{(\alpha-1)(\beta-1)}{\alpha+\beta-3}}}{\alpha + \beta - 2}$$

- ($1 < \alpha < 2$, $0 < \beta < 1$) The distribution has a mode at the right end $x=1$ and it is negatively skewed, left-tailed. There is **one inflection point**, located to the left of the mode:

$$x = \frac{\alpha - 1 - \sqrt{\frac{(\alpha-1)(\beta-1)}{\alpha+\beta-3}}}{\alpha + \beta - 2}$$

There are no inflection points in the remaining (symmetric and skewed) regions: U-shaped: ($\alpha, \beta < 1$) upside-down-U-shaped: ($1 < \alpha < 2$, $1 < \beta < 2$), reverse-J-shaped ($\alpha < 1$, $\beta > 2$) or J-shaped: ($\alpha > 2$, $\beta < 1$)

The accompanying plots show the inflection point locations (shown vertically, ranging from 0 to 1) versus α and β (the horizontal axes ranging from 0 to 5). There are large cuts at surfaces intersecting the lines $\alpha = 1$, $\beta = 1$, $\alpha = 2$, and $\beta = 2$ because at these values the beta distribution change from 2 modes, to 1 mode to no mode.

Shapes

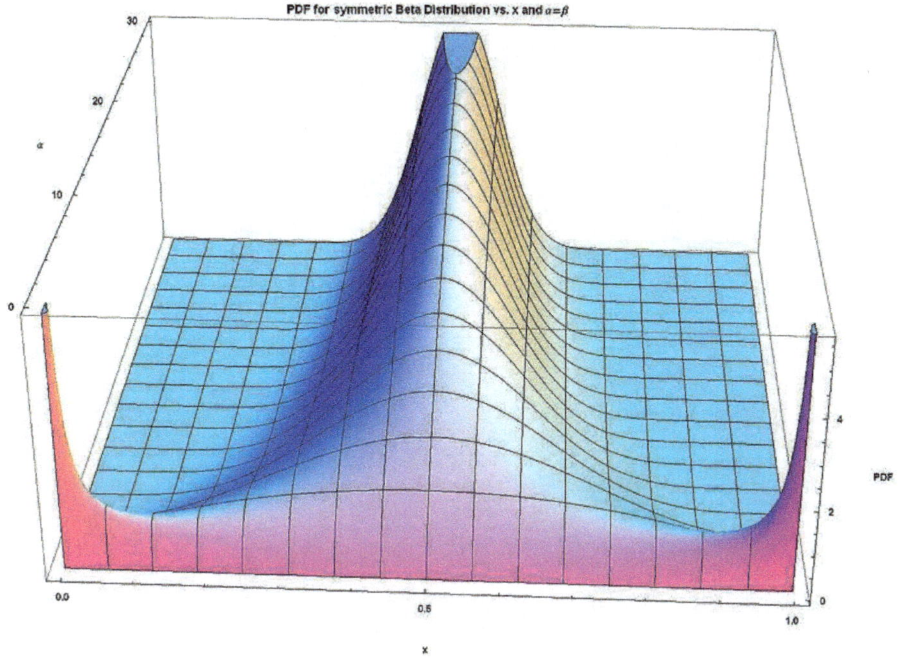

PDF for symmetric beta distribution vs. x and alpha=beta from 0 to 30

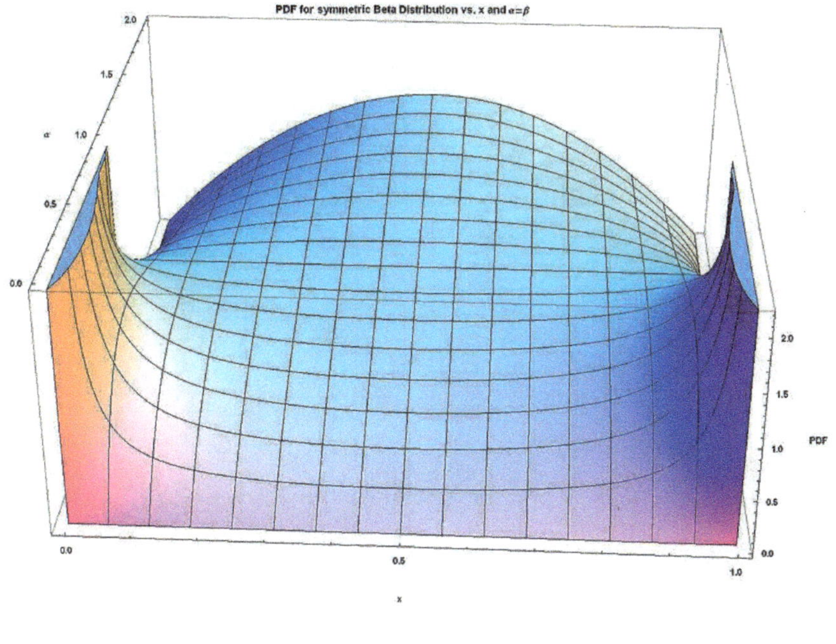

PDF for symmetric beta distribution vs. x and alpha=beta from 0 to 2

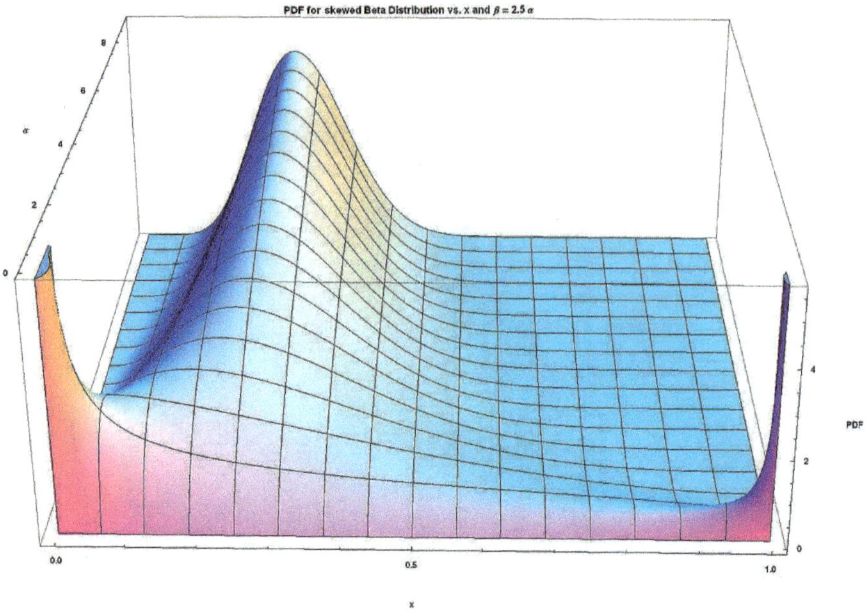

PDF for skewed beta distribution vs. x and beta= 2.5 alpha from 0 to 9

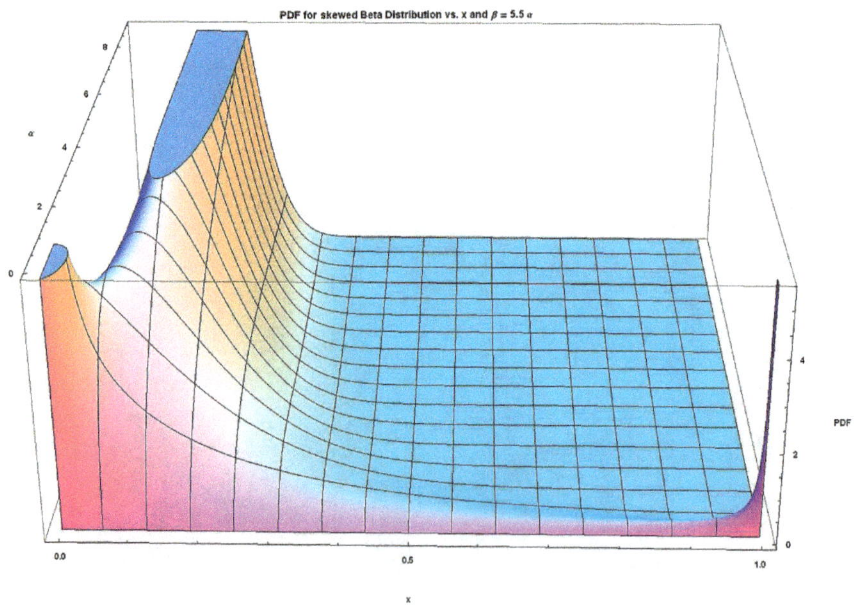

PDF for skewed beta distribution vs. x and beta= 5.5 alpha from 0 to 9

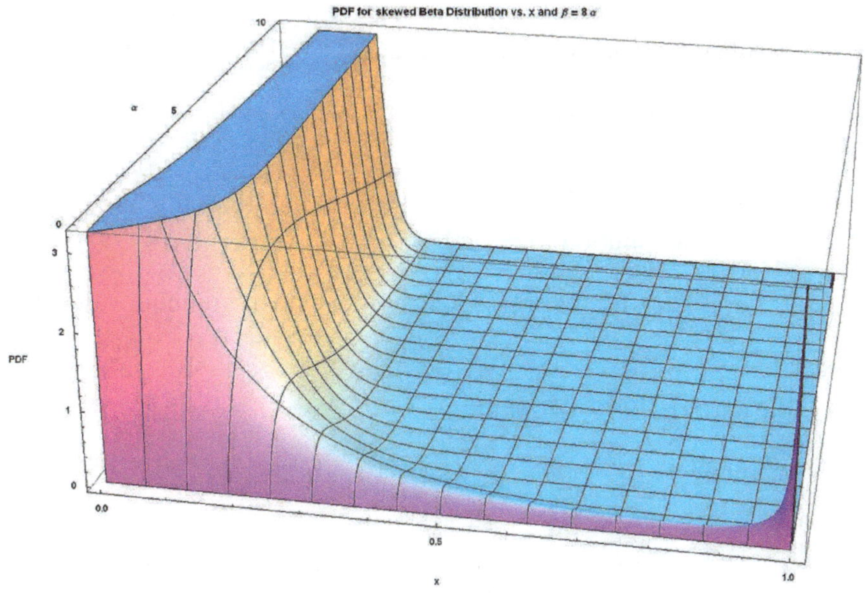

PDF for skewed beta distribution vs. x and beta= 8 alpha from 0 to 10.

The beta density function can take a wide variety of different shapes depending on the values of the two parameters α and β. The ability of the beta distribution to take this great diversity of shapes (using only two parameters) is partly responsible for finding wide application for modeling actual measurements:

Symmetric (α = β)

- the density function is symmetric about 1/2 (blue & teal plots).
- median = mean = 1/2.
- skewness = 0.
- **α = β < 1**
    - U-shaped (blue plot).
    - bimodal: left mode = 0, right mode =1, anti-mode = 1/2
    - 1/12 < var(X) < 1/4
    - −2 < excess kurtosis(X) < −6/5
    - α = β = 1/2 is the arcsine distribution
        - var(X) = 1/8
        - excess kurtosis(X) = −3/2
    - α = β → 0 is a 2-point Bernoulli distribution with equal probability 1/2 at each

Dirac delta function end x = 0 and x = 1 and zero probability everywhere else. A coin toss: one face of the coin being x = 0 and the other face being x = 1.

- $\lim_{\alpha=\beta\to 0} \operatorname{var}(X) = \frac{1}{4}$
- $\lim_{\alpha=\beta\to 0} \text{excess kurtosis}(X) = -2$ a lower value than this is impossible for any distribution to reach.
- The differential entropy approaches a minimum value of $-\infty$

■ α = β = 1

- the uniform [0, 1] distribution
- no mode
- var(X) = 1/12
- excess kurtosis(X) = −6/5
- The (negative anywhere else) differential entropy reaches its maximum value of zero

■ α = β > 1

- symmetric unimodal
- mode = 1/2.
- 0 < var(X) < 1/12
- −6/5 < excess kurtosis(X) < 0
- α = β = 3/2 is a semi-elliptic [0, 1] distribution, see: Wigner semicircle distribution
    - var(X) = 1/16.
    - excess kurtosis(X) = −1
- α = β = 2 is the parabolic [0, 1] distribution
    - var(X) = 1/20
    - excess kurtosis(X) = −6/7
- α = β > 2 is bell-shaped, with inflection points located to either side of the mode
    - 0 < var(X) < 1/20

- $-6/7 <$ excess kurtosis(X) < 0

- $\alpha = \beta \rightarrow \infty$ is a 1-point Degenerate distribution with a Dirac delta function spike at the midpoint x = 1/2 with probability 1, and zero probability everywhere else. There is 100% probability (absolute certainty) concentrated at the single point x = 1/2.

    - $\lim_{\alpha=\beta\to\infty} \mathrm{var}(X) = 0$

    - $\lim_{\alpha=\beta\to\infty} \text{excess kurtosis}(X) = 0$

    - The differential entropy approaches a minimum value of $-\infty$

Skewed (α ≠ β)

The density function is skewed. An interchange of parameter values yields the mirror image (the reverse) of the initial curve, some more specific cases:

- $\alpha < 1, \beta < 1$
    - U-shaped
    - Positive skew for $\alpha < \beta$, negative skew for $\alpha > \beta$.
    - bimodal: left mode = 0, right mode = 1, anti-mode = $\frac{\alpha-1}{\alpha+\beta-2}$,
    - $0 <$ median < 1.
    - $0 < \mathrm{var}(X) < 1/4$
- $\alpha > 1, \beta > 1$
    - unimodal (magenta & cyan plots),
    - Positive skew for $\alpha < \beta$, negative skew for $\alpha > \beta$.
    - $\frac{\alpha-1}{\alpha+\beta-2}$
    - $0 <$ median < 1
    - $0 < \mathrm{var}(X) < 1/12$
- $\alpha < 1, \beta \geq 1$
    - reverse J-shaped with a right tail,
    - positively skewed,
    - strictly decreasing, convex
    - mode = 0

- $0 <$ median $< 1/2$.
- $0 < \mathrm{var}(X) < \frac{-11+5\sqrt{5}}{2}$, (maximum variance occurs for $\alpha = 1, \beta = \frac{-1+\sqrt{5}}{2}$, or β = Φ the golden ratio conjugate)

■ $\alpha \geq 1,\ \beta < 1$

- J-shaped with a left tail,
- negatively skewed,
- strictly increasing, convex
- mode = 1
- $1/2 <$ median < 1
- $0 < \mathrm{var}(X) < \frac{-11+5\sqrt{5}}{2}$, (maximum variance occurs for $\alpha = 1, \beta = \frac{-1+\sqrt{5}}{2}$, or β = Φ the golden ratio conjugate)

■ $\alpha = 1,\ \beta > 1$

- positively skewed,
- strictly decreasing (red plot),
- a reversed (mirror-image) power function [0,1] distribution
- mode = 0
- $\alpha = 1,\ 1 < \beta < 2$
    - concave
    - $1 - \frac{1}{\sqrt{2}} < \text{median} < \frac{1}{2}$
    - $1/18 < \mathrm{var}(X) < 1/12$.
- $\alpha = 1,\ \beta = 2$
    - a straight line with slope −2, the right-triangular distribution with right angle at the left end, at x = 0
    - $\text{median} = 1 - \frac{1}{\sqrt{2}}$
    - $\mathrm{var}(X) = 1/18$
- $\alpha = 1,\ \beta > 2$
    - reverse J-shaped with a right tail,

- convex
- $0 < \text{median} < 1 - \frac{1}{\sqrt{2}}$
- 0 < var(X) < 1/18

■ α > 1, β = 1

- negatively skewed,
- strictly increasing (green plot),
- the power function [0, 1] distribution
- mode =1
- 2 > α > 1, β = 1
    - concave
    - $0 < \text{median} < 1 - \frac{1}{\sqrt{2}}$
    - 1/18 < var(X) < 1/12
- α = 2, β = 1
    - a straight line with slope +2, the right-triangular distribution with right angle at the right end, at x = 1
    - $\frac{1}{2} < \text{median} < \frac{1}{\sqrt{2}}$
    - var(X) = 1/18
- α > 2, β = 1
    - J-shaped with a left tail, convex
    - $\frac{1}{\sqrt{2}} < \text{median} < 1$
    - 0 < var(X) < 1/18

Parameter Estimation

Method of Moments

Two Unknown Parameters

Two unknown parameters $((\hat{\alpha}, \hat{\beta})$ of a beta distribution supported in the [0,1] interval) can be estimated, using the method of moments, with the first two moments (sample mean and sample variance) as follows. Let:

$$\text{sample mean(X)} = \bar{x} = \frac{1}{N}\sum_{i=1}^{N} X_i$$

be the sample mean estimate and

$$\text{sample variance(X)} = \bar{v} = \frac{1}{N-1}\sum_{i=1}^{N}(X_i - \bar{x})^2$$

be the sample variance estimate. The method-of-moments estimates of the parameters are

$$\hat{\alpha} = \bar{x}\left(\frac{\bar{x}(1-\bar{x})}{\bar{v}} - 1\right), \text{ if } \bar{v} < \bar{x}(1-\bar{x}),$$

$$\hat{\beta} = (1-\bar{x})\left(\frac{\bar{x}(1-\bar{x})}{\bar{v}} - 1\right), \text{ if } \bar{v} < \bar{x}(1-\bar{x}).$$

When the distribution is required over known interval other than [0, 1] with random variable X, say [a, c] with random variable Y, then replace $\bar{x}$ with $\frac{\bar{y}-a}{c-a}$, and $\bar{v}$ with $\frac{\bar{v_Y}}{(c-a)^2}$ in the above couple of equations for the shape parameters (see "Alternative parametrizations, four parameters" section below)., where:

$$\text{sample mean(Y)} = \bar{y} = \frac{1}{N}\sum_{i=1}^{N} Y_i$$

$$\text{sample variance(Y)} = \bar{v_Y} = \frac{1}{N-1}\sum_{i=1}^{N}(Y_i - \bar{y})^2$$

Four Unknown Parameters

Solutions for parameter estimates vs. (sample) excess Kurtosis and (sample) squared Skewness Beta distribution

All four parameters $(\hat{\alpha}, \hat{\beta}, \hat{a}, \hat{c}$ of a beta distribution supported in the [a, c] interval -see section "Alternative parametrizations, Four parameters"-) can be estimated, using the method of moments developed by Karl Pearson, by equating sample and population values of the first four central moments (mean, variance, skewness and excess kurtosis). The excess kurtosis was expressed in terms of the square of the skewness, and the sample size ν = α + β, (see previous section "Kurtosis") as follows:

$$\text{excess kurtosis} = \frac{6}{3+\nu}\left(\frac{(2+\nu)}{4}(\text{skewness})^2 - 1\right) \text{ if (skewness)}^2 - 2 < \text{excess kurtosis} < \tfrac{3}{2}(\text{skewness})^2$$

One can use this equation to solve for the sample size ν= α + β in terms of the square of the skewness and the excess kurtosis as follows:

$$\hat{\nu} = \hat{\alpha} + \hat{\beta} = 3\frac{(\text{sample excess kurtosis}) - (\text{sample skewness})^2 + 2}{\frac{3}{2}(\text{sample skewness})^2 - (\text{sample excess kurtosis})}$$

$$\text{if (sample skewness)}^2 - 2 < \text{sample excess kurtosis} < \tfrac{3}{2}(\text{sample skewness})^2$$

This is the ratio (multiplied by a factor of 3) between the previously derived limit boundaries for the beta distribution in a space (as originally done by Karl Pearson) defined with coordinates of the square of the skewness in one axis and the excess kurtosis in the other axis (see previous section titled "Kurtosis bounded by the square of the skewness"):

The case of zero skewness, can be immediately solved because for zero skewness, $\alpha = \beta$ and hence $\nu = 2\alpha = 2\beta$, therefore $\alpha = \beta = \nu/2$

$$\hat{\alpha} = \hat{\beta} = \frac{\hat{\nu}}{2} = \frac{\frac{3}{2}(\text{sample excess kurtosis}) + 3}{-(\text{sample excess kurtosis})}$$
$$\text{if sample skewness} = 0 \text{ and } -2 < \text{sample excess kurtosis} < 0$$

(Excess kurtosis is negative for the beta distribution with zero skewness, ranging from -2 to 0, so that $\hat{\nu}$ -and therefore the sample shape parameters- is positive, ranging from zero when the shape parameters approach zero and the excess kurtosis approaches -2, to infinity when the shape parameters approach infinity and the excess kurtosis approaches zero).

For non-zero sample skewness one needs to solve a system of two coupled equations. Since the skewness and the excess kurtosis are independent of the parameters $\hat{a}, \hat{c}$, the parameters $\hat{\alpha}, \hat{\beta}$ can be uniquely determined from the sample skewness and the sample excess kurtosis, by solving the coupled equations with two known variables (sample skewness and sample excess kurtosis) and two unknowns (the shape parameters):

$$(\text{sample skewness})^2 = \frac{4(\hat{\beta} - \hat{\alpha})^2(1 + \hat{\alpha} + \hat{\beta})}{\hat{\alpha}\hat{\beta}(2 + \hat{\alpha} + \hat{\beta})^2}$$
$$\text{sample excess kurtosis} = \frac{6}{3 + \hat{\alpha} + \hat{\beta}}\left(\frac{(2 + \hat{\alpha} + \hat{\beta})}{4}(\text{sample skewness})^2 - 1\right)$$
$$\text{if } (\text{sample skewness})^2 - 2 < \text{sample excess kurtosis} < \tfrac{3}{2}(\text{sample skewness})^2$$

resulting in the following solution:

$$\hat{\alpha}, \hat{\beta} = \frac{\hat{\nu}}{2}\left(1 \pm \frac{1}{\sqrt{1 + \frac{16(\hat{\nu}+1)}{(\hat{\nu}+2)^2(\text{sample skewness})^2}}}\right)$$
$$\text{if sample skewness} \neq 0 \text{ and } (\text{sample skewness})^2 - 2 < \text{sample excess kurtosis} < \tfrac{3}{2}(\text{sample skewness})^2$$

Where one should take the solutions as follows: $\hat{\alpha} > \hat{\beta}$ for (negative) sample skewness < 0, and $\hat{\alpha} < \hat{\beta}$ for (positive) sample skewness > 0.

The accompanying plot shows these two solutions as surfaces in a space with horizontal axes of (sample excess kurtosis) and (sample squared skewness) and the shape parameters as the vertical axis. The surfaces are constrained by the condition that the sample excess kurtosis must be bounded by the sample squared skewness as stipulated in the above equation. The two surfaces meet at the right edge defined by zero skewness. Along this right edge, both parameters are equal and the distribution is symmetric U-shaped for $\alpha = \beta < 1$, uniform for $\alpha = \beta = 1$, upside-down-U-shaped for $1 < \alpha = \beta < 2$ and bell-shaped for $\alpha = \beta > 2$. The surfaces also meet at the front (lower) edge defined by the impossible boundary line (excess kurtosis + 2 - skewness^2 = 0). Along this front (lower) boundary both shape parameters approach zero, and the probability density is concentrated more at one end than the other end (with practically nothing in between), with probabilities $p = \frac{\beta}{\alpha+\beta}$ at the left end x = 0 and $q = 1 - p = \frac{\alpha}{\alpha+\beta}$ at the right end x = 1. The two surfaces become further apart towards the rear edge. At this rear edge the surface parameters are quite different from each other. As remarked, for example, by Bowman and Shenton, sampling in the neighborhood of the line (sample excess kurtosis - (3/2)(sample skewness)^2 = 0) (the just-J-shaped portion of the rear edge where blue meets beige), "is dangerously near to chaos", because at that line the denominator of the expression above for the estimate ν = α + β becomes zero and hence ν approaches infinity as that line is approached. Bowman and Shenton write that the higher moment parameters (kurtosis and skewness) are extremely fragile (near that line). However the mean and standard deviation are fairly reliable. Therefore, the problem is for the case of four parameter estimation for very skewed distributions such that the excess kurtosis approaches (3/2) times the square of the skewness. This boundary line is produced by extremely skewed distributions with very large values of one of the parameters and very small values of the other parameter. See section titled Kurtosis bounded by the square of the skewness for a numerical example and further comments about this rear edge boundary line (sample excess kurtosis - (3/2) (sample skewness)^2 = 0). As remarked by Karl Pearson himself this issue may not be of much practical importance as this trouble arises only for very skewed J-shaped (or mirror-image J-shaped) distributions with very different values of shape parameters that are unlikely to occur much in practice). The usual skewed skewed-bell-shape distributions that occur in practice do not have this parameter estimation problem.

The remaining two parameters $\hat{a}, \hat{c}$ can be determined using the sample mean and the sample variance using a variety of equations. One alternative is to calculate the support interval range $(\hat{c} - \hat{a})$ based on the sample variance and the sample kurtosis. For this purpose one can solve, in terms of the range $(\hat{c} - \hat{a})$, the equation expressing the excess kurtosis in terms of the sample variance, and the sample size ν (see section titled "Kurtosis" and "Alternative parametrizations, four parameters"):

$$\text{sample excess kurtosis} = \frac{6}{(3+\hat{\nu})(2+\hat{\nu})}\left(\frac{(\hat{c}-\hat{a})^2}{(\text{sample variance})} - 6 - 5\hat{\nu}\right)$$

to obtain:

$$(\hat{c}-\hat{a}) = \sqrt{(\text{sample variance})}\sqrt{6 + 5\hat{\nu} + \frac{(2+\hat{\nu})(3+\hat{\nu})}{6}(\text{sample excess kurtosis})}$$

Another alternative is to calculate the support interval range $(\hat{c} - \hat{a})$ based on the sample variance and the sample skewness. For this purpose one can solve, in terms of the range $(\hat{c} - \hat{a})$, the equation expressing the squared skewness in terms of the sample variance, and the sample size ν (see section titled "Skewness" and "Alternative parametrizations, four parameters"):

$$(\text{sample skewness})^2 = \frac{4}{(2+\hat{\nu})^2}\left(\frac{(\hat{c}-\hat{a})^2}{(\text{sample variance})} - 4(1+\hat{\nu})\right)$$

to obtain:

$$(\hat{c} - \hat{a}) = \frac{\sqrt{(\text{sample variance})}}{2}\sqrt{(2+\hat{\nu})^2(\text{sample skewness})^2 + 16(1+\hat{\nu})}$$

The remaining parameter can be determined from the sample mean and the previously obtained parameters: $(\hat{c} - \hat{a}), \hat{\alpha}, \hat{\nu} = \hat{\alpha} + \hat{\beta}$:

$$\hat{a} = (\text{sample mean}) - \left(\frac{\hat{\alpha}}{\hat{\nu}}\right)(\hat{c} - \hat{a})$$

and finally, of course, $\hat{c} = (\hat{c} - \hat{a}) + \hat{a}$.

In the above formulas one may take, for example, as estimates of the sample moments:

$$\text{sample mean} = \overline{y} = \frac{1}{N}\sum_{i=1}^{N} Y_i$$

$$\text{sample variance} = \overline{v}_Y = \frac{1}{N-1}\sum_{i=1}^{N}(Y_i - \overline{y})^2$$

$$\text{sample skewness} = G_1 = \frac{N}{(N-1)(N-2)}\frac{\sum_{i=1}^{N}(Y_i - \overline{y})^3}{\overline{v}_Y^{\frac{3}{2}}}$$

$$\text{sample excess kurtosis} = G_2 = \frac{N(N+1)}{(N-1)(N-2)(N-3)}\frac{\sum_{i=1}^{N}(Y_i - \overline{y})^4}{\overline{v}_Y^2} - \frac{3(N-1)^2}{(N-2)(N-3)}$$

The estimators G_1 for sample skewness and G_2 for sample kurtosis are used by DAP/SAS, PSPP/SPSS, and Excel. However, they are not used by BMDP and (according to) they were not used by MINITAB in 1998. Actually, Joanes and Gill in their 1998 study concluded that the skewness and kurtosis estimators used in BMDP and in MINITAB (at that time) had smaller variance and mean-squared error in normal samples, but the skewness and kurtosis estimators used in DAP/SAS, PSPP/SPSS, namely G_1 and G_2, had smaller mean-squared error in samples from a very skewed distribution. It is for this reason that we have spelled out "sample skewness", etc., in the above formulas, to make it explicit that the user should choose the best estimator according to the problem at hand, as the best estimator for skewness and kurtosis depends on the amount of skewness (as shown by Joanes and Gill).

Maximum likelihood

Two Unknown Parameters

Max (Joint Log Likelihood/N) for Beta distribution Maxima at alpha=beta=2

Max (Joint Log Likelihood/N) for Beta distribution Maxima at alpha=beta={0.25,0.5,1,2,4,6,8}

As it is also the case for maximum likelihood estimates for the gamma distribution, the maximum likelihood estimates for the beta distribution do not have a general closed form solution for arbitrary values of the shape parameters. If $X_1, ..., X_N$ are independent random variables each having a beta distribution, the joint log likelihood function for N iid observations is:

$$\begin{aligned}\ln \mathcal{L}(\alpha,\beta|X) &= \sum_{i=1}^{N} \ln(\mathcal{L}_i(\alpha,\beta|X_i)) \\ &= \sum_{i=1}^{N} \ln(f(X_i;\alpha,\beta)) \\ &= \sum_{i=1}^{N} \ln\left(\frac{X_i^{\alpha-1}(1-X_i)^{\beta-1}}{\mathrm{B}(\alpha,\beta)}\right) \\ &= (\alpha-1)\sum_{i=1}^{N}\ln(X_i) + (\beta-1)\sum_{i=1}^{N}\ln(1-X_i) - N\ln \mathrm{B}(\alpha,\beta)\end{aligned}$$

Finding the maximum with respect to a shape parameter involves taking the partial derivative with respect to the shape parameter and setting the expression equal to zero yielding the maximum likelihood estimator of the shape parameters:

$$\frac{\partial \ln \mathcal{L}(\alpha,\beta|X)}{\partial\alpha} = \sum_{i=1}^{N}\ln X_i - N\frac{\partial \ln \mathrm{B}(\alpha,\beta)}{\partial\alpha} = 0$$
$$\frac{\partial \ln \mathcal{L}(\alpha,\beta|X)}{\partial\beta} = \sum_{i=1}^{N}\ln(1-X_i) - N\frac{\partial \ln \mathrm{B}(\alpha,\beta)}{\partial\beta} = 0$$

where:

$$\frac{\partial \ln \mathrm{B}(\alpha,\beta)}{\partial\alpha} = -\frac{\partial \ln\Gamma(\alpha+\beta)}{\partial\alpha} + \frac{\partial \ln\Gamma(\alpha)}{\partial\alpha} + \frac{\partial \ln\Gamma(\beta)}{\partial\alpha} = -\psi(\alpha+\beta)+\psi(\alpha)+0$$
$$\frac{\partial \ln \mathrm{B}(\alpha,\beta)}{\partial\beta} = -\frac{\partial \ln\Gamma(\alpha+\beta)}{\partial\beta} + \frac{\partial \ln\Gamma(\alpha)}{\partial\beta} + \frac{\partial \ln\Gamma(\beta)}{\partial\beta} = -\psi(\alpha+\beta)+0+\psi(\beta)$$

since the **digamma function** denoted $\psi(\alpha)$ is defined as the logarithmic derivative of the gamma function:

$$\psi(\alpha) = \frac{\partial \ln\Gamma(\alpha)}{\partial\alpha}$$

To ensure that the values with zero tangent slope are indeed a maximum (instead of a saddle-point or a minimum) one has to also satisfy the condition that the curvature is negative. This amounts to satisfying that the second partial derivative with respect to the shape parameters is negative

$$\frac{\partial^2 \ln \mathcal{L}(\alpha,\beta|X)}{\partial\alpha^2} = -N\frac{\partial^2 \ln \mathrm{B}(\alpha,\beta)}{\partial\alpha^2} < 0$$
$$\frac{\partial^2 \ln \mathcal{L}(\alpha,\beta|X)}{\partial\beta^2} = -N\frac{\partial^2 \ln \mathrm{B}(\alpha,\beta)}{\partial\beta^2} < 0$$

using the previous equations, this is equivalent to:

$$\frac{\partial^2 \ln \mathrm{B}(\alpha,\beta)}{\partial\alpha^2} = \psi_1(\alpha) - \psi_1(\alpha+\beta) > 0$$
$$\frac{\partial^2 \ln \mathrm{B}(\alpha,\beta)}{\partial\beta^2} = \psi_1(\beta) - \psi_1(\alpha+\beta) > 0$$

where the **trigamma function**, denoted $\psi_1(\alpha)$, is the second of the polygamma functions, and is defined as the derivative of the digamma function:

$$\psi_1(\alpha) = \frac{\partial^2 \ln \Gamma(\alpha)}{\partial \alpha^2} = \frac{\partial \psi(\alpha)}{\partial \alpha}.$$

These conditions are equivalent to stating that the variances of the logarithmically transformed variables are positive, since:

$$\mathrm{var}[\ln(X)] = \mathrm{E}[\ln^2(X)] - (\mathrm{E}[\ln(X)])^2 = \psi_1(\alpha) - \psi_1(\alpha+\beta)$$
$$\mathrm{var}[\ln(1-X)] = \mathrm{E}[\ln^2(1-X)] - (\mathrm{E}[\ln(1-X)])^2 = \psi_1(\beta) - \psi_1(\alpha+\beta)$$

Therefore, the condition of negative curvature at a maximum is equivalent to the statements:

$$\mathrm{var}[\ln(X)] > 0$$
$$\mathrm{var}[\ln(1-X)] > 0$$

Alternatively, the condition of negative curvature at a maximum is also equivalent to stating that the following logarithmic derivatives of the geometric means G_X and $G_{(1-X)}$ are positive, since:

$$\psi_1(\alpha) - \psi_1(\alpha+\beta) = \frac{\partial \ln G_X}{\partial \alpha} > 0$$
$$\psi_1(\beta) - \psi_1(\alpha+\beta) = \frac{\partial \ln G_{(1-X)}}{\partial \beta} > 0$$

While these slopes are indeed positive, the other slopes are negative:

$$\frac{\partial \ln G_X}{\partial \beta}, \frac{\partial \ln G_{(1-X)}}{\partial \alpha} < 0.$$

The slopes of the mean and the median with respect to α and β display similar sign behavior.

From the condition that at a maximum, the partial derivative with respect to the shape parameter equals zero, we obtain the following system of coupled maximum likelihood estimate equations (for the average log-likelihoods) that needs to be inverted to obtain the (unknown) shape parameter estimates $\hat{\alpha}, \hat{\beta}$ in terms of the (known) average of logarithms of the samples $X_1, \ldots, X_N$:

$$\hat{\mathrm{E}}[\ln(X)] = \psi(\hat{\alpha}) - \psi(\hat{\alpha}+\hat{\beta}) = \frac{1}{N}\sum_{i=1}^{N} \ln X_i = \ln \hat{G}_X$$

$$\hat{\mathrm{E}}[\ln(1-X)] = \psi(\hat{\beta}) - \psi(\hat{\alpha}+\hat{\beta}) = \frac{1}{N}\sum_{i=1}^{N} \ln(1-X_i) = \ln \hat{G}_{(1-X)}$$

where we recognize $\log \hat{G}_X$ as the logarithm of the sample geometric mean and $\log \hat{G}_{(1-X)}$ as the logarithm of the sample geometric mean based on (1-X), the mirror-image of X. For $\hat{\alpha} = \hat{\beta}$, it follows that $\hat{G}_X = \hat{G}_{(1-X)}$.

$$\hat{G}_X = \prod_{i=1}^{N} (X_i)^{\frac{1}{N}}$$
$$\hat{G}_{(1-X)} = \prod_{i=1}^{N} (1-X_i)^{\frac{1}{N}}$$

These coupled equations containing digamma functions of the shape parameter estimates $\hat{\alpha}, \hat{\beta}$ must be solved by numerical methods as done, for example, by Beckman et al. Gnanadesikan et al. give numerical solutions for a few cases.N.L.Johnson and S.Kotz suggest that for "not too small" shape parameter estimates $\hat{\alpha}, \hat{\beta}$, the logarithmic approximation to the digamma function $\psi(\hat{\alpha}) \approx \ln(\hat{\alpha} - \tfrac{1}{2})$ may be used to obtain initial values for an iterative solution, since the equations resulting from this approximation can be solved exactly:

$$\ln \frac{\hat{\alpha} - \frac{1}{2}}{\hat{\alpha} + \hat{\beta} - \frac{1}{2}} \approx \ln \hat{G}_X$$

$$\ln \frac{\hat{\beta} - \frac{1}{2}}{\hat{\alpha} + \hat{\beta} - \frac{1}{2}} \approx \ln \hat{G}_{(1-X)}$$

which leads to the following solution for the initial values (of the estimate shape parameters in terms of the sample geometric means) for an iterative solution:

$$\hat{\alpha} \approx \tfrac{1}{2} + \frac{\hat{G}_X}{2(1 - \hat{G}_X - \hat{G}_{(1-X)})} \text{ if } \hat{\alpha} > 1$$

$$\hat{\beta} \approx \tfrac{1}{2} + \frac{\hat{G}_{(1-X)}}{2(1 - \hat{G}_X - \hat{G}_{(1-X)})} \text{ if } \hat{\beta} > 1$$

Alternatively, the estimates provided by the method of moments can instead be used as initial values for an iterative solution of the maximum likelihood coupled equations in terms of the digamma functions.

When the distribution is required over a known interval other than [0, 1] with random variable X, say [a, c] with random variable Y, then replace $\ln(X_i)$ in the first equation with

$$\ln \frac{Y_i - a}{c - a},$$

and replace $\ln(1-X_i)$ in the second equation with

$$\ln \frac{c - Y_i}{c - a}$$

(see "Alternative parametrizations, four parameters" section below).

If one of the shape parameters is known, the problem is considerably simplified. The following logit transformation can be used to solve for the unknown shape parameter (for skewed cases such that $\hat{\alpha} \neq \hat{\beta}$, otherwise, if symmetric, both -equal- parameters are known when one is known):

$$\hat{\mathrm{E}}\left[\ln\left(\frac{X}{1-X}\right)\right] = \psi(\hat{\alpha}) - \psi(\hat{\beta}) = \frac{1}{N}\sum_{i=1}^{N} \ln \frac{X_i}{1 - X_i} = \ln \hat{G}_X - \ln\left(\hat{G}_{(1-X)}\right)$$

This logit transformation is the logarithm of the transformation that divides the variable X by its mirror-image (X/(1 - X) resulting in the "inverted beta distribution" or beta prime distribution (also known as beta distribution of the second kind or Pearson›s Type VI) with support [0, +∞). As previously discussed in the section Moments of logarithmically transformed random variables, the logit transformation $\frac{X}{1-X}$, studied by Johnson, extends the finite support [0, 1] based on the original variable X to infinite support in both directions of the real line (−∞, +∞).

If, for example, $\hat{\beta}$ is known, the unknown parameter $\hat{\alpha}$ can be obtained in terms of the inverse digamma function of the right hand side of this equation:

$$\psi(\hat{\alpha}) = \frac{1}{N}\sum_{i=1}^{N} \ln \frac{X_i}{1-X_i} + \psi(\hat{\beta})$$

$$\hat{\alpha} = (\textbf{Inverse digamma})[\ln \hat{G}_X - \ln \hat{G}_{(1-X)} + \psi(\hat{\beta})]$$

In particular, if one of the shape parameters has a value of unity, for example for $\hat{\beta} = 1$ (the power function distribution with bounded support [0,1]), using the identity ψ(x + 1) = ψ(x) + 1/x in the equation ψ(x + 1) = ψ(x) + 1/x in the equation $\psi(\hat{\alpha}) - \psi(\hat{\alpha} + \hat{\beta}) = \ln \hat{G}_X$, the maximum likelihood estimator for the unknown parameter $\hat{\alpha}$ is, exactly:

$$\hat{\alpha} = -\frac{1}{\frac{1}{N}\sum_{i=1}^{N} \ln X_i} = -\frac{1}{\ln \hat{G}_X}$$

The beta has support [0, 1], therefore $\hat{G}_X < 1$, and hence $(-\ln \hat{G}_X) > 0$, and therefore $\hat{\alpha} > 0$.

In conclusion, the maximum likelihood estimates of the shape parameters of a beta distribution are (in general) a complicated function of the sample geometric mean, and of the sample geometric mean based on (1−X), the mirror-image of X. One may ask, if the variance (in addition to the mean) is necessary to estimate two shape parameters with the method of moments, why is the (logarithmic or geometric) variance not necessary to estimate two shape parameters with the maximum likelihood method, for which only the geometric means suffice? The answer is because the mean does not provide as much information as the geometric mean. For a beta distribution with equal shape parameters α = β, the mean is exactly 1/2, regardless of the value of the shape parameters, and therefore regardless of the value of the statistical dispersion (the variance). On the other hand, the geometric mean of a beta distribution with equal shape parameters α = β, depends on the value of the shape parameters, and therefore it contains more information. Also, the geometric mean of a beta distribution does not satisfy the symmetry conditions satisfied by the mean, therefore, by employing both the geometric mean based on X and geometric mean based on (1−X), the maximum likelihood method is able to provide best estimates for both parameters α = β, without need of employing the variance.

One can express the joint log likelihood per N iid observations in terms of the sufficient statistics (the sample geometric means) as follows:

$$\frac{\ln \mathcal{L}(\alpha, \beta|X)}{N} = (\alpha - 1)\ln \hat{G}_X + (\beta - 1)\ln \hat{G}_{(1-X)} - \ln \mathrm{B}(\alpha, \beta)$$

We can plot the joint log likelihood per N observations for fixed values of the sample geometric means to see the behavior of the likelihood function as a function of the shape parameters α and β. In such a plot, the shape parameter estimators $\hat{\alpha}, \hat{\beta}$

correspond to the maxima of the likelihood function. See the accompanying graph that shows that all the likelihood functions intersect at $\alpha = \beta = 1$, which corresponds to the values of the shape parameters that give the maximum entropy (the maximum entropy occurs for shape parameters equal to unity: the uniform distribution). It is evident from the plot that the likelihood function gives sharp peaks for values of the shape parameter estimators close to zero, but that for values of the shape parameters estimators greater than one, the likelihood function becomes quite flat, with less defined peaks. Obviously, the maximum likelihood parameter estimation method for the beta distribution becomes less acceptable for larger values of the shape parameter estimators, as the uncertainty in the peak definition increases with the value of the shape parameter estimators. One can arrive at the same conclusion by noticing that the expression for the curvature of the likelihood function is in terms of the geometric variances

$$\frac{\partial^2 \ln \mathcal{L}(\alpha,\beta|X)}{\partial\alpha^2} = -\operatorname{var}[\ln X]$$
$$\frac{\partial^2 \ln \mathcal{L}(\alpha,\beta|X)}{\partial\beta^2} = -\operatorname{var}[\ln(1-X)]$$

These variances (and therefore the curvatures) are much larger for small values of the shape parameter α and β. However, for shape parameter values $\alpha, \beta > 1$, the variances (and therefore the curvatures) flatten out. Equivalently, this result follows from the Cramér–Rao bound, since the Fisher information matrix components for the beta distribution are these logarithmic variances. The Cramér–Rao bound states that the variance of any unbiased estimator $\hat{\alpha}$ of α is bounded by the reciprocal of the Fisher information:

$$\frac{\partial^2 \ln \mathcal{L}(\alpha,\beta|X)}{\partial\alpha^2} = -\operatorname{var}[\ln X]$$
$$\frac{\partial^2 \ln \mathcal{L}(\alpha,\beta|X)}{\partial\beta^2} = -\operatorname{var}[\ln(1-X)]$$

so the variance of the estimators increases with increasing α and β, as the logarithmic variances decrease.

Also one can express the joint log likelihood per N iid observations in terms of the digamma function expressions for the logarithms of the sample geometric means as follows:

$$\operatorname{var}(\hat{\alpha}) \geq \frac{1}{\operatorname{var}[\ln X]} \geq \frac{1}{\psi_1(\hat{\alpha}) - \psi_1(\hat{\alpha}+\hat{\beta})}$$
$$\operatorname{var}(\hat{\beta}) \geq \frac{1}{\operatorname{var}[\ln(1-X)]} \geq \frac{1}{\psi_1(\hat{\beta}) - \psi_1(\hat{\alpha}+\hat{\beta})}$$

this expression is identical to the negative of the cross-entropy (see section on "Quantities of information (entropy)"). Therefore, finding the maximum of the joint log likelihood of the shape parameters, per N iid observations, is identical to finding the minimum of the cross-entropy for the beta distribution, as a function of the shape parameters.

$$\frac{\ln \mathcal{L}(\alpha,\beta|X)}{N} = -H = -h - D_{\mathrm{KL}} = -\ln \mathrm{B}(\alpha,\beta) + (\alpha-1)\psi(\hat{\alpha}) + (\beta-1)\psi(\hat{\beta}) - (\alpha+\beta-2)\psi(\hat{\alpha}+$$

with the cross-entropy defined as follows:

$$H = \int_0^1 -f(X;\hat{\alpha},\hat{\beta})\ln(f(X;\alpha,\beta))\,\mathrm{d}X$$

Four Unknown Parameters

The procedure is similar to the one followed in the two unknown parameter case. If $Y_1, \ldots, Y_N$ are independent random variables each having a beta distribution with four parameters, the joint log likelihood function for N iid observations is:

$$\begin{aligned}
\ln \mathcal{L}(\alpha,\beta,a,c|Y) &= \sum_{i=1}^{N} \ln \mathcal{L}_i(\alpha,\beta,a,c|Y_i) \\
&= \sum_{i=1}^{N} \ln f(Y_i;\alpha,\beta,a,c) \\
&= \sum_{i=1}^{N} \ln \frac{(Y_i-a)^{\alpha-1}(c-Y_i)^{\beta-1}}{(c-a)^{\alpha+\beta-1}\mathrm{B}(\alpha,\beta)} \\
&= (\alpha-1)\sum_{i=1}^{N}\ln(Y_i-a) + (\beta-1)\sum_{i=1}^{N}\ln(c-Y_i) - N\ln \mathrm{B}(\alpha,\beta) - N(\alpha+\beta-1)\ln(c-a)
\end{aligned}$$

Finding the maximum with respect to a shape parameter involves taking the partial derivative with respect to the shape parameter and setting the expression equal to zero yielding the maximum likelihood estimator of the shape parameters:

$$\frac{\partial \ln \mathcal{L}(\alpha,\beta,a,c|Y)}{\partial \alpha} = \sum_{i=1}^{N}\ln(Y_i-a) - N(-\psi(\alpha+\beta)+\psi(\alpha)) - N\ln(c-a) = 0$$

$$\frac{\partial \ln \mathcal{L}(\alpha,\beta,a,c|Y)}{\partial \beta} = \sum_{i=1}^{N}\ln(c-Y_i) - N(-\psi(\alpha+\beta)+\psi(\beta)) - N\ln(c-a) = 0$$

$$\frac{\partial \ln \mathcal{L}(\alpha,\beta,a,c|Y)}{\partial a} = -(\alpha-1)\sum_{i=1}^{N}\frac{1}{Y_i-a} + N(\alpha+\beta-1)\frac{1}{c-a} = 0$$

$$\frac{\partial \ln \mathcal{L}(\alpha,\beta,a,c|Y)}{\partial c} = (\beta-1)\sum_{i=1}^{N}\frac{1}{c-Y_i} - N(\alpha+\beta-1)\frac{1}{c-a} = 0$$

these equations can be re-arranged as the following system of four coupled equations (the first two equations are geometric means and the second two equations are the harmonic means) in terms of the maximum likelihood estimates for the four parameters $\hat{\alpha},\hat{\beta},\hat{a},\hat{c}$:

$$\frac{1}{N}\sum_{i=1}^{N}\ln\frac{Y_i-\hat{a}}{\hat{c}-\hat{a}} = \psi(\hat{\alpha}) - \psi(\hat{\alpha}+\hat{\beta}) = \ln \hat{G}_X$$

$$\frac{1}{N}\sum_{i=1}^{N}\ln\frac{\hat{c}-Y_i}{\hat{c}-\hat{a}} = \psi(\hat{\beta}) - \psi(\hat{\alpha}+\hat{\beta}) = \ln \hat{G}_{1-X}$$

$$\frac{1}{\frac{1}{N}\sum_{i=1}^{N}\frac{\hat{c}-\hat{a}}{Y_i-\hat{a}}} = \frac{\hat{\alpha}-1}{\hat{\alpha}+\hat{\beta}-1} = \hat{H}_X$$

$$\frac{1}{\frac{1}{N}\sum_{i=1}^{N}\frac{\hat{c}-\hat{a}}{\hat{c}-Y_i}} = \frac{\hat{\beta}-1}{\hat{\alpha}+\hat{\beta}-1} = \hat{H}_{1-X}$$

with sample geometric means:

$$\hat{G}_X = \prod_{i=1}^{N}\left(\frac{Y_i-\hat{a}}{\hat{c}-\hat{a}}\right)^{\frac{1}{N}}$$

$$\hat{G}_{(1-X)} = \prod_{i=1}^{N}\left(\frac{\hat{c}-Y_i}{\hat{c}-\hat{a}}\right)^{\frac{1}{N}}$$

The parameters $\hat{a},\hat{c}$ are embedded inside the geometric mean expressions in a nonlinear way (to the power 1/N). This precludes, in general, a closed form solution, even for an initial value approximation for iteration purposes. One alternative is to use as initial values for iteration the values obtained from the method of moments solution for the four parameter case. Furthermore, the expressions for the harmonic means are well-defined only for $\hat{\alpha}, \hat{\beta} > 1$, which precludes a maximum likelihood solution for shape parameters less than unity in the four-parameter case. Fisher's information matrix for the four parameter case is positive-definite only for α, β > 2 (for further discussion, see section on Fisher information matrix, four parameter case), for bell-shaped (symmetric or unsymmetric) beta distributions, with inflection points located to either side of the mode. The following Fisher information components (that represent the expectations of the curvature of the log likelihood function) have singularities at the following values:

$$\alpha = 2: \quad \mathrm{E}\left[-\frac{1}{N}\frac{\partial^2 \ln \mathcal{L}(\alpha,\beta,a,c|Y)}{\partial a^2}\right] = \mathcal{I}_{a,a}$$

$$\beta = 2: \quad \mathrm{E}\left[-\frac{1}{N}\frac{\partial^2 \ln \mathcal{L}(\alpha,\beta,a,c|Y)}{\partial c^2}\right] = \mathcal{I}_{c,c}$$

$$\alpha = 2: \quad \mathrm{E}\left[-\frac{1}{N}\frac{\partial^2 \ln \mathcal{L}(\alpha,\beta,a,c|Y)}{\partial \alpha \partial a}\right] = \mathcal{I}_{\alpha,a}$$

$$\beta = 1: \quad \mathrm{E}\left[-\frac{1}{N}\frac{\partial^2 \ln \mathcal{L}(\alpha,\beta,a,c|Y)}{\partial \beta \partial c}\right] = \mathcal{I}_{\beta,c}$$

(for further discussion see section on Fisher information matrix). Thus, it is not possible to strictly carry on the maximum likelihood estimation for some well known distributions belonging to the four-parameter beta distribution family, like the uniform distribution (Beta(1, 1, a, c)), and the arcsine distribution (Beta(1/2, 1/2, a, c)). N.L.Johnson and S.Kotz ignore the equations for the harmonic means and instead suggest "If a and c are unknown, and maximum likelihood estimators of a, c, α and β are required, the above procedure (for the two unknown parameter case, with X transformed as X = (Y−a)/(c−a)) can be repeated using a succession of trial values of a and c, until the pair (a, c) for which maximum likelihood (given a and c) is as great as possible, is attained" (where, for the purpose of clarity, their notation for the parameters has been translated into the present notation).

Fisher Information Matrix

Let a random variable X have a probability density f(x;α). The partial derivative with respect to the (unknown, and to be estimated) parameter α of the log likelihood function is called the score. The second moment of the score is called the Fisher information:

$$\mathcal{I}(\alpha) = \mathrm{E}\left[\left(\frac{\partial}{\partial \alpha} \ln \mathcal{L}(\alpha|X)\right)^2\right],$$

The expectation of the score is zero, therefore the Fisher information is also the second moment centered on the mean of the score: the variance of the score.

If the log likelihood function is twice differentiable with respect to the parameter α, and under certain regularity conditions,then the Fisher information may also be written as follows (which is often a more convenient form for calculation purposes):

$$\text{var}[\hat{\alpha}] \geq \frac{1}{\mathcal{I}(\alpha)}.$$

Thus, the Fisher information is the negative of the expectation of the second derivative with respect to the parameter α of the log likelihood function. Therefore, Fisher information is a measure of the curvature of the log likelihood function of α. A low curvature (and therefore high radius of curvature), flatter log likelihood function curve has low Fisher information; while a log likelihood function curve with large curvature (and therefore low radius of curvature) has high Fisher information. When the Fisher information matrix is computed at the evaluates of the parameters (the observed Fisher information matrix) it is equivalent to the replacement of the true log likelihood surface by a Taylor›s series approximation, taken as far as the quadratic terms. The word information, in the context of Fisher information, refers to information about the parameters. Information such as: estimation, sufficiency and properties of variances of estimators. The Cramér–Rao bound states that the inverse of the Fisher information is a lower bound on the variance of any estimator of a parameter α:

$$\text{var}[\hat{\alpha}] \geq \frac{1}{\mathcal{I}(\alpha)}.$$

The precision to which one can estimate the estimator of a parameter α is limited by the Fisher Information of the log likelihood function. The Fisher information is a measure of the minimum error involved in estimating a parameter of a distribution and it can be viewed as a measure of the resolving power of an experiment needed to discriminate between two alternative hypothesis of a parameter.

When there are N parameters

$$\begin{bmatrix} \theta_1 \\ \theta_2 \\ \dots \\ \theta_N \end{bmatrix},$$

then the Fisher information takes the form of an N×N positive semidefinite symmetric matrix, the Fisher Information Matrix, with typical element:

$$(\mathcal{I}(\theta))_{i,j} = \mathrm{E}\left[\left(\frac{\partial}{\partial\theta_i}\ln\mathcal{L}\right)\left(\frac{\partial}{\partial\theta_j}\ln\mathcal{L}\right)\right]$$

Under certain regularity conditions, the Fisher Information Matrix may also be written in the following form, which is often more convenient for computation:

$$(\mathcal{I}(\theta))_{i,j} = -\mathrm{E}\left[\frac{\partial^2}{\partial\theta_i\,\partial\theta_j}\ln(\mathcal{L})\right].$$

With X_1, ..., X_N iid random variables, an N-dimensional “box” can be constructed with sides X_1, ..., X_N. Costa and Covershow that the (Shannon) differential entropy h(X) is related to the volume of the typical set (having the sample entropy close to the true entropy), while the Fisher information is related to the surface of this typical set.

Two Parameters

For $X_1, \ldots, X_N$ independent random variables each having a beta distribution parametrized with shape parameters α and β, the joint log likelihood function for N iid observations is:

$$\ln(\mathcal{L}(\alpha,\beta|X)) = (\alpha-1)\sum_{i=1}^{N}\ln X_i + (\beta-1)\sum_{i=1}^{N}\ln(1-X_i) - N\ln \mathrm{B}(\alpha,\beta)$$

therefore the joint log likelihood function per N iid observations is:

$$\frac{1}{N}\ln(\mathcal{L}(\alpha,\beta|X)) = (\alpha-1)\frac{1}{N}\sum_{i=1}^{N}\ln X_i + (\beta-1)\frac{1}{N}\sum_{i=1}^{N}\ln(1-X_i) - \ln \mathrm{B}(\alpha,\beta)$$

For the two parameter case, the Fisher information has 4 components: 2 diagonal and 2 off-diagonal. Since the Fisher information matrix is symmetric, one of these off diagonal components is independent. Therefore, the Fisher information matrix has 3 independent components (2 diagonal and 1 off diagonal).

Aryal and Nadarajah calculated Fisher's information matrix for the four parameter case, from which the two parameter case can be obtained as follows:

$$-\frac{\partial^2 \ln \mathcal{L}(\alpha,\beta|X)}{N\partial\alpha^2} = \mathrm{var}[\ln(X)] = \psi_1(\alpha) - \psi_1(\alpha+\beta) = \mathcal{I}_{\alpha,\alpha} = \mathrm{E}\left[-\frac{\partial^2 \ln \mathcal{L}(\alpha,\beta|X)}{N\partial\alpha^2}\right] = \ln \mathrm{var}_{GX}$$

$$-\frac{\partial^2 \ln \mathcal{L}(\alpha,\beta|X)}{N\partial\beta^2} = \mathrm{var}[\ln(1-X)] = \psi_1(\beta) - \psi_1(\alpha+\beta) = \mathcal{I}_{\beta,\beta} = \mathrm{E}\left[-\frac{\partial^2 \ln \mathcal{L}(\alpha,\beta|X)}{N\partial\beta^2}\right] = \ln \mathrm{var}_{G(1-X)}$$

$$-\frac{\partial^2 \ln \mathcal{L}(\alpha,\beta|X)}{N\partial\alpha\partial\beta} = \mathrm{cov}[\ln X, \ln(1-X)] = -\psi_1(\alpha+\beta) = \mathcal{I}_{\alpha,\beta} = \mathrm{E}\left[-\frac{\partial^2 \ln \mathcal{L}(\alpha,\beta|X)}{N\partial\alpha\partial\beta}\right] = \ln \mathrm{cov}_{GX,(1-X)}$$

Since the Fisher information matrix is symmetric

$$\mathcal{I}_{\alpha,\beta} = \mathcal{I}_{\beta,\alpha} = \ln \mathrm{cov}_{GX,(1-X)}$$

The Fisher information components are equal to the log geometric variances and log geometric covariance. Therefore, they can be expressed as **trigamma functions**, denoted $\psi_1(\alpha)$, the second of the polygamma functions, defined as the derivative of the digamma function:

$$\psi_1(\alpha) = \frac{d^2 \ln\Gamma(\alpha)}{\partial\alpha^2} = \frac{\partial\psi(\alpha)}{\partial\alpha}.$$

These derivatives are also derived in the section titled "Parameter estimation", "Maximum likelihood", "Two unknown parameters," and plots of the log likelihood function are also shown in that section. The section titled "Geometric variance and covariance" contains plots and further discussion of the Fisher information matrix components: the log geometric variances and log geometric covariance as a function of the shape parameters α and β. The section titled "Other moments", "Moments of transformed random variables", "Moments of logarithmically transformed random variables" contains formulas for moments of logarithmically transformed random variables. Images for the Fisher information components $\mathcal{I}_{\alpha,\alpha}$, $\mathcal{I}_{\beta,\beta}$ and $\mathcal{I}_{\alpha,\beta}$ are shown in the section titled "Geometric variance".

The determinant of Fisher's information matrix is of interest (for example for the calculation of Jeffreys prior probability). From the expressions for the individual components of the Fisher information matrix, it follows that the determinant of Fisher's (symmetric) information matrix for the beta distribution is:

$$\begin{aligned}\det(\mathcal{I}(\alpha,\beta)) &= \mathcal{I}_{\alpha,\alpha}\mathcal{I}_{\beta,\beta} - \mathcal{I}_{\alpha,\beta}\mathcal{I}_{\alpha,\beta}\\ &= (\psi_1(\alpha)-\psi_1(\alpha+\beta))(\psi_1(\beta)-\psi_1(\alpha+\beta)) - (-\psi_1(\alpha+\beta))(-\psi_1(\alpha+\beta))\\ &= \psi_1(\alpha)\psi_1(\beta) - (\psi_1(\alpha)+\psi_1(\beta))\psi_1(\alpha+\beta)\\ \lim_{\alpha\to 0}\det(\mathcal{I}(\alpha,\beta)) &= \lim_{\beta\to 0}\det(\mathcal{I}(\alpha,\beta)) = \infty\\ \lim_{\alpha\to \infty}\det(\mathcal{I}(\alpha,\beta)) &= \lim_{\beta\to \infty}\det(\mathcal{I}(\alpha,\beta)) = 0\end{aligned}$$

From Sylvester's criterion (checking whether the diagonal elements are all positive), it follows that the Fisher information matrix for the two parameter case is positive-definite (under the standard condition that the shape parameters are positive $\alpha>0$ and $\beta>0$).

Four Parameters

Fisher Information I(a,a) for $\alpha=\beta$ vs range (c-a) and exponent $\alpha=\beta$

Fisher Information I(α,a) for $\alpha=\beta$, vs. range (c - a) and exponent $\alpha=\beta$

If $Y_1, ..., Y_N$ are independent random variables each having a beta distribution with four parameters: the exponents α and β, as well as "a" (the minimum of the distribution range), and "c" (the maximum of the distribution range) (section titled "Alternative parametrizations", "Four parameters"), with probability density function:

$$f(y;\alpha,\beta,a,c) = \frac{f(x;\alpha,\beta)}{c-a} = \frac{\left(\frac{y-a}{c-a}\right)^{\alpha-1}\left(\frac{c-y}{c-a}\right)^{\beta-1}}{(c-a)B(\alpha,\beta)} = \frac{(y-a)^{\alpha-1}(c-y)^{\beta-1}}{(c-a)^{\alpha+\beta-1}B(\alpha,\beta)}.$$

the joint log likelihood function per N iid observations is:

$$\frac{1}{N}\ln(\mathcal{L}(\alpha,\beta,a,c|Y)) = \frac{\alpha-1}{N}\sum_{i=1}^{N}\ln(Y_i-a) + \frac{\beta-1}{N}\sum_{i=1}^{N}\ln(c-Y_i) - \ln \mathrm{B}(\alpha,\beta) - (\alpha+\beta-1)\ln(c-a)$$

For the four parameter case, the Fisher information has 4*4=16 components. It has 12 off-diagonal components = (4*4 total - 4 diagonal). Since the Fisher information matrix is symmetric, half of these components (12/2=6) are independent. Therefore, the Fisher information matrix has 6 independent off-diagonal + 4 diagonal = 10 independent components. Aryal and Nadarajahcalculated Fisher's information matrix for the four parameter case as follows:

$$-\frac{1}{N}\frac{\partial^2\ln\mathcal{L}(\alpha,\beta,a,c|Y)}{\partial\alpha^2} = \mathrm{var}[\ln(X)] = \psi_1(\alpha)-\psi_1(\alpha+\beta) = \mathcal{I}_{\alpha,\alpha} = \mathrm{E}\left[-\frac{1}{N}\frac{\partial^2\ln\mathcal{L}(\alpha,\beta,a,c|Y)}{\partial\alpha^2}\right] = \ln(\mathrm{var_{GX}})$$

$$-\frac{1}{N}\frac{\partial^2\ln\mathcal{L}(\alpha,\beta,a,c|Y)}{\partial\beta^2} = \mathrm{var}[\ln(1-X)] = \psi_1(\beta)-\psi_1(\alpha+\beta) = \mathcal{I}_{\beta,\beta} = \mathrm{E}\left[-\frac{1}{N}\frac{\partial^2\ln\mathcal{L}(\alpha,\beta,a,c|Y)}{\partial\beta^2}\right] = \ln(\mathrm{var_{G(1-X)}})$$

$$-\frac{1}{N}\frac{\partial^2\ln\mathcal{L}(\alpha,\beta,a,c|Y)}{\partial\alpha\partial\beta} = \mathrm{cov}[\ln X,(1-X)] = -\psi_1(\alpha+\beta) = \mathcal{I}_{\alpha,\beta} = \mathrm{E}\left[-\frac{1}{N}\frac{\partial^2\ln\mathcal{L}(\alpha,\beta,a,c|Y)}{\partial\alpha\partial\beta}\right] = \ln(\mathrm{cov_{GX,(1-X)}})$$

In the above expressions, the use of X instead of Y in the expressions var[ln(X)] = ln(var_{GX}) is not an error. The expressions in terms of the log geometric variances and log geometric covariance occur as functions of the two parameter X ~ Beta(α, β) parametrization because when taking the partial derivatives with respect to the exponents (α, β) in the four parameter case, one obtains the identical expressions as for the two parameter case: these terms of the four parameter Fisher information matrix are independent of the minimum "a" and maximum "c" of the distribution's range. The only non-zero term upon double differentiation of the log likelihood function with respect to the exponents α and β is the second derivative of the log of the beta function: ln(B(α, β)). This term is independent of the minimum a and maximum c of the distribution›s range. Double differentiation of this term results in trigamma functions. The sections titled Maximum likelihood, Two unknown parameters and Four unknown parameters also show this fact.

The Fisher information for N i.i.d. samples is N times the individual Fisher information (eq. 11.279, page 394 of Cover and Thomas). (Aryal and Nadarajah take a single observation, N = 1, to calculate the following components of the Fisher information, which leads to the same result as considering the derivatives of the log likelihood per N observations. Moreover, below the erroneous expression for $\mathcal{I}_{a,a}$ in Aryal and Nadarajah has been corrected.)

$$\alpha > 2: \quad \mathrm{E}\left[-\frac{1}{N}\frac{\partial^2 \ln \mathcal{L}(\alpha,\beta,a,c|Y)}{\partial a^2}\right] = \mathcal{I}_{a,a} = \frac{\beta(\alpha+\beta-1)}{(\alpha-2)(c-a)^2}$$

$$\beta > 2: \quad \mathrm{E}\left[-\frac{1}{N}\frac{\partial^2 \ln \mathcal{L}(\alpha,\beta,a,c|Y)}{\partial c^2}\right] = \mathcal{I}_{c,c} = \frac{\alpha(\alpha+\beta-1)}{(\beta-2)(c-a)^2}$$

$$\mathrm{E}\left[-\frac{1}{N}\frac{\partial^2 \ln \mathcal{L}(\alpha,\beta,a,c|Y)}{\partial a\,\partial c}\right] = \mathcal{I}_{a,c} = \frac{(\alpha+\beta-1)}{(c-a)^2}$$

$$\alpha > 1: \quad \mathrm{E}\left[-\frac{1}{N}\frac{\partial^2 \ln \mathcal{L}(\alpha,\beta,a,c|Y)}{\partial \alpha\,\partial a}\right] = \mathcal{I}_{\alpha,a} = \frac{\beta}{(\alpha-1)(c-a)}$$

$$\mathrm{E}\left[-\frac{1}{N}\frac{\partial^2 \ln \mathcal{L}(\alpha,\beta,a,c|Y)}{\partial \alpha\,\partial c}\right] = \mathcal{I}_{\alpha,c} = \frac{1}{(c-a)}$$

$$\mathrm{E}\left[-\frac{1}{N}\frac{\partial^2 \ln \mathcal{L}(\alpha,\beta,a,c|Y)}{\partial \beta\,\partial a}\right] = \mathcal{I}_{\beta,a} = -\frac{1}{(c-a)}$$

$$\beta > 1: \quad \mathrm{E}\left[-\frac{1}{N}\frac{\partial^2 \ln \mathcal{L}(\alpha,\beta,a,c|Y)}{\partial \beta\,\partial c}\right] = \mathcal{I}_{\beta,c} = -\frac{\alpha}{(\beta-1)(c-a)}$$

The lower two diagonal entries of the Fisher information matrix, with respect to the parameter "a" (the minimum of the distribution's range): $\mathcal{I}_{a,a}$, and with respect to the parameter "c" (the maximum of the distribution's range): $\mathcal{I}_{c,c}$ are only defined for exponents α > 2 and β > 2 respectively. The Fisher information matrix component $\mathcal{I}_{a,a}$ for the minimum "a approaches infinity for exponent α approaching 2 from above, and the Fisher information matrix component $\mathcal{I}_{c,c}$ for the maximum "c" approaches infinity for exponent β approaching 2 from above.

The Fisher information matrix for the four parameter case does not depend on the individual values of the minimum a and the maximum c, but only on the total range (c−a). Moreover, the components of the Fisher information matrix that depend on the range (c−a), depend only through its inverse (or the square of the inverse), such that the Fisher information decreases for increasing range (c−a).

The accompanying images show the Fisher information components $\mathcal{I}_{a,a}$ and $\mathcal{I}_{\alpha,a}$. Images for the Fisher information components $\mathcal{I}_{\alpha,\alpha}$ and $\mathcal{I}_{\beta,\beta}$ are shown in the section titled "Geometric variance". All these Fisher information components look like a basin, with the "walls" of the basin being located at low values of the parameters.

The following four-parameter-beta-distribution Fisher information components can be expressed in terms of the two-parameter: X ~ Beta(α, β) expectations of the transformed ratio ((1-X)/X) and of its mirror image (X/(1-X)), scaled by the range (c−a), which may be helpful for interpretation:

$$\mathcal{I}_{\alpha,a} = \frac{\mathrm{E}\left[\frac{1-X}{X}\right]}{c-a} = \frac{\beta}{(\alpha-1)(c-a)} \text{ if } \alpha > 1$$

$$\mathcal{I}_{\beta,c} = -\frac{\mathrm{E}\left[\frac{X}{1-X}\right]}{c-a} = -\frac{\alpha}{(\beta-1)(c-a)} \text{ if } \beta > 1$$

These are also the expected values of the "inverted beta distribution" or beta prime distribution (also known as beta distribution of the second kind or Pearson›s Type VI) and its mirror image, scaled by the range (c−a).

Also, the following Fisher information components can be expressed in terms of the harmonic (1/X) variances or of variances based on the ratio transformed variables ((1-X)/X) as follows:

$$\alpha > 2: \quad \mathcal{I}_{a,a} = \operatorname{var}\left[\frac{1}{X}\right]\left(\frac{\alpha-1}{c-a}\right)^2 = \operatorname{var}\left[\frac{1-X}{X}\right]\left(\frac{\alpha-1}{c-a}\right)^2 = \frac{\beta(\alpha+\beta-1)}{(\alpha-2)(c-a)^2}$$

$$\beta > 2: \quad \mathcal{I}_{c,c} = \operatorname{var}\left[\frac{1}{1-X}\right]\left(\frac{\beta-1}{c-a}\right)^2 = \operatorname{var}\left[\frac{X}{1-X}\right]\left(\frac{\beta-1}{c-a}\right)^2 = \frac{\alpha(\alpha+\beta-1)}{(\beta-2)(c-a)^2}$$

$$\mathcal{I}_{a,c} = \operatorname{cov}\left[\frac{1}{X},\frac{1}{1-X}\right]\frac{(\alpha-1)(\beta-1)}{(c-a)^2} = \operatorname{cov}\left[\frac{1-X}{X},\frac{X}{1-X}\right]\frac{(\alpha-1)(\beta-1)}{(c-a)^2} = \frac{(\alpha+\beta-1)}{(c-a)^2}$$

See section "Moments of linearly transformed, product and inverted random variables" for these expectations.

The determinant of Fisher's information matrix is of interest (for example for the calculation of Jeffreys prior probability). From the expressions for the individual components, it follows that the determinant of Fisher's (symmetric) information matrix for the beta distribution with four parameters is:

$$\begin{aligned}\det(\mathcal{I}(\alpha,\beta,a,c)) = & -\mathcal{I}_{a,c}^2\mathcal{I}_{\alpha,a}\mathcal{I}_{\alpha,\beta} + \mathcal{I}_{a,a}\mathcal{I}_{a,c}\mathcal{I}_{\alpha,c}\mathcal{I}_{\alpha,\beta} + \mathcal{I}_{a,c}^2\mathcal{I}_{\alpha,\beta}^2 - \mathcal{I}_{a,a}\mathcal{I}_{c,c}\mathcal{I}_{\alpha,\beta}^2 \\ & - \mathcal{I}_{a,c}\mathcal{I}_{\alpha,a}\mathcal{I}_{\alpha,c}\mathcal{I}_{\beta,a} + \mathcal{I}_{a,c}^2\mathcal{I}_{\alpha,\alpha}\mathcal{I}_{\beta,a} + 2\mathcal{I}_{c,c}\mathcal{I}_{\alpha,a}\mathcal{I}_{\alpha,\beta}\mathcal{I}_{\beta,a} \\ & - 2\mathcal{I}_{a,c}\mathcal{I}_{\alpha,c}\mathcal{I}_{\alpha,\beta}\mathcal{I}_{\beta,a} + \mathcal{I}_{\alpha,c}^2\mathcal{I}_{\beta,a}^2 - \mathcal{I}_{c,c}\mathcal{I}_{\alpha,\alpha}\mathcal{I}_{\beta,a}^2 + \mathcal{I}_{a,c}\mathcal{I}_{\alpha,a}^2\mathcal{I}_{\beta,c} \\ & - \mathcal{I}_{a,a}\mathcal{I}_{a,c}\mathcal{I}_{\alpha,\alpha}\mathcal{I}_{\beta,c} - \mathcal{I}_{a,c}\mathcal{I}_{\alpha,a}\mathcal{I}_{\alpha,\beta}\mathcal{I}_{\beta,c} + \mathcal{I}_{a,a}\mathcal{I}_{\alpha,c}\mathcal{I}_{\alpha,\beta}\mathcal{I}_{\beta,c} \\ & - \mathcal{I}_{\alpha,a}\mathcal{I}_{\alpha,c}\mathcal{I}_{\beta,a}\mathcal{I}_{\beta,c} + \mathcal{I}_{a,c}\mathcal{I}_{\alpha,\alpha}\mathcal{I}_{\beta,a}\mathcal{I}_{\beta,c} - \mathcal{I}_{c,c}\mathcal{I}_{\alpha,a}^2\mathcal{I}_{\beta,\beta} \\ & + 2\mathcal{I}_{a,c}\mathcal{I}_{\alpha,a}\mathcal{I}_{\alpha,c}\mathcal{I}_{\beta,\beta} - \mathcal{I}_{a,a}\mathcal{I}_{\alpha,c}^2\mathcal{I}_{\beta,\beta} - \mathcal{I}_{a,c}^2\mathcal{I}_{\alpha,\alpha}\mathcal{I}_{\beta,\beta} + \mathcal{I}_{a,a}\mathcal{I}_{c,c}\mathcal{I}_{\alpha,\alpha}\mathcal{I}_{\beta,\beta} \text{ if } \alpha,\beta > 2\end{aligned}$$

Using Sylvester's criterion (checking whether the diagonal elements are all positive), and since diagonal components $\mathcal{I}_{a,a}$ and $\mathcal{I}_{c,c}$ have singularities at α=2 and β=2 it follows that the Fisher information matrix for the four parameter case is positive-definite for α>2 and β>2. Since for α > 2 and β > 2 the beta distribution is (symmetric or unsymmetric) bell shaped, it follows that the Fisher information matrix is positive-definite only for bell-shaped (symmetric or unsymmetric) beta distributions, with inflection points located to either side of the mode. Thus, important well known distributions belonging to the four-parameter beta distribution family, like the parabolic distribution (Beta(2,2,a,c)) and the uniform distribution (Beta(1,1,a,c)) have Fisher information components $(\mathcal{I}_{a,a}, \mathcal{I}_{c,c}, \mathcal{I}_{\alpha,a}, \mathcal{I}_{\beta,c})$ that blow up (approach infinity) in the four-parameter case (although their Fisher information components are all defined for the two parameter case). The four-parameter Wigner semicircle distribution (Beta(3/2,3/2,a,c)) and arcsine distribution (Beta(1/2,1/2,a,c)) have negative Fisher information determinants for the four-parameter case.

Generating beta-distributed random variates

If X and Y are independent, with $X \sim \Gamma(\alpha, \theta)$ and $Y \sim \Gamma(\beta, \theta)$ then

$\frac{X}{X+Y} \sim \mathrm{B}(\alpha, \beta)$.

So one algorithm for generating beta variates is to generate X/(X + Y), where X is a gamma variate with parameters (α, 1) and Y is an independent gamma variate with parameters (β, 1).

Also, the kth order statistic of n uniformly distributed variates is $\mathrm{B}(k, n+1-k)$, so an alternative if α and β are small integers is to generate α + β − 1 uniform variates and choose the α-th smallest.

Another way to generate the Beta distribution is by Pólya urn model. According to this method, one start with an “urn” with α “black” balls and β “white” balls and draw uniformly with replacement. Every trial an additional ball is added according to the color of the last ball which was drawn. Asymptotically, the proportion of black and white balls will be distributed according to the Beta distribution, where each repetition of the experiment will produce a different value.

Related Distributions

Transformations

- If X ~ Beta(α, β) then 1−X ~ Beta(β, α) mirror-image symmetry
- If X ~ Beta(α, β) then $\frac{X}{1-X} \sim \beta'(\alpha, \beta)$. The beta prime distribution, also called “beta distribution of the second kind”.
- If X ~ Beta(n/2, m/2) then $\frac{mX}{n(1-X)} \sim F(n, m)$ (assuming n > 0 and m > 0), the Fisher-Snedecor F distribution.
- If $X \sim Beta\left(1 + \lambda\frac{m-\min}{\max-\min}, 1 + \lambda\frac{\max-m}{\max-\min}\right)$ then min+X(max−min) ~ PERT(min, max, m, λ) where PERT denotes a distribution used in PERT analysis, and m=most likely value. Traditionally λ = 4 in PERT analysis.
- If X ~ Beta(1, β) then X ~ Kumaraswamy distribution with parameters (1, β)
- If X ~ Beta(α, 1) then X ~ Kumaraswamy distribution with parameters (α, 1)
- If X ~ Beta(α, 1) then −ln(X) ~ Exponential(α)

Special and Limiting Cases

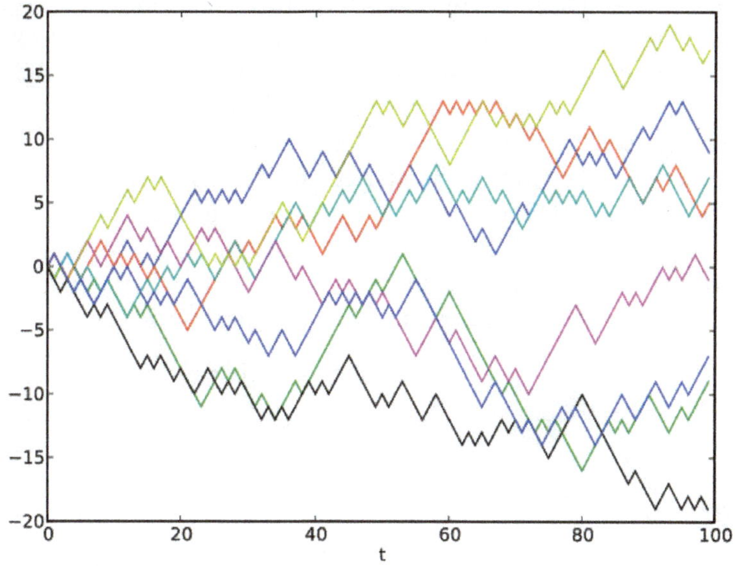

Example of eight realizations of a random walk in one dimension starting at 0: the probability for the time of the last visit to the origin is distributed as Beta(1/2, 1/2).

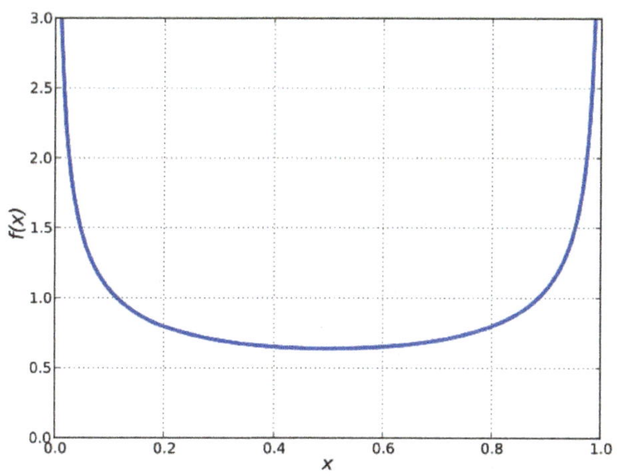

Beta(1/2, 1/2): The arcsine distribution probability density was proposed by Harold Jeffreys to represent uncertainty for a Bernoulli or a binomial distribution in Bayesian inference, and is now commonly referred to as Jeffreys prior: $p^{-1/2}(1-p)^{-1/2}$. This distribution also appears in several random walk fundamental theorems.

- Beta(1, 1) ~ U(0, 1).
- If X ~ Beta(3/2, 3/2) and r > 0 then 2rX−r ~ Wigner semicircle distribution.
- Beta(1/2, 1/2) is equivalent to the arcsine distribution. This distribution is also Jeffreys prior probability for the Bernoulli and binomial distributions . The arcsine probability density is a distribution that appears in several random walk fundamental theorems. In a fair coin toss random walk, the probability for the time of the last visit to the origin is distributed as an (U-shaped) arcsine distribution. In a two-player fair-coin-toss game, a player is said to be in the lead if the random walk (that started at the origin) is above the origin. The most probable number of times that a given player will be in the lead, in a game of length 2N, is not N. On the contrary, N is the least likely number of times that the player will be in the lead. The most likely number of times in the lead is 0 or 2N (following the arcsine distribution).
- $\lim_{n\to\infty} nBeta(1,n) = \text{Exponential}(1)$ the exponential distribution
- $\lim_{n\to\infty} nBeta(k,n) = \text{Gamma}(k,1)$ the gamma distribution

Derived from other Distributions

- The kth order statistic of a sample of size n from the uniform distribution is a beta random variable, $U_{(k)}$ ~ Beta(k, n+1−k).
- If X ~ Gamma(α, θ) and Y ~ Gamma(β, θ) are independent, then $\frac{X}{X+Y} \sim Beta(\alpha,\beta)$.
- If $X \sim \chi^2(\alpha)$ and $Y \sim \chi^2(\beta)$ are independent, then $\frac{X}{X+Y} \sim Beta(\frac{\alpha}{2}, \frac{\beta}{2})$.
- If X ~ U(0, 1) and α > 0 then $X^{1/\alpha}$ ~ Beta(α, 1). The power function distribution.

Combination with other Distributions

- X ~ Beta(α, β) and Y ~ F(2β,2α) then {\displaystyle \Pr(X\leq {\tfrac {\alpha }{\alpha +\beta x}})=\Pr(Y\geq x)\,} for all x > 0.

Compounding with other distributions

- If p ~ Beta(α, β) and X ~ Bin(k, p) then X ~ beta-binomial distribution
- If p ~ Beta(α, β) and X ~ NB(r, p) then X ~ beta negative binomial distribution

Generalisations

- The Dirichlet distribution is a multivariate generalization of the beta distribution. Univariate marginals of the Dirichlet distribution have a beta distribution. The beta distribution is conjugate to the binomial and Bernoulli distributions in exactly the same way as the Dirichlet distribution is conjugate to the multinomial distribution and categorical distribution.

- The Pearson type I distribution is identical to the beta distribution (except for arbitrary shifting and re-scaling that can also be accomplished with the four parameter parametrization of the beta distribution).
- $Beta(\alpha, \beta) = \lim_{\delta \to 0} \text{NonCentralBeta}(\alpha, \beta, \delta)$ the noncentral beta distribution
- the Generalized beta distribution is a five-parameter distribution family which has the beta distribution as a special case.

REFERENCES

1. Pierre Simon de Laplace (1812). Analytical Theory of Probability.
2. The first major treatise blending calculus with probability theory, originally in French: Théorie Analytique des Probabilités.
3. A. Kolmogoroff (1933). Grundbegriffe der Wahrscheinlichkeitsrechnung. doi:10.1007/978-3-642-49888-6. ISBN 978-3-642-49888-6.
4. An English translation by Nathan Morrison appeared under the title Foundations of the Theory of Probability (Chelsea, New York) in 1950, with a second edition in 1956.
5. Patrick Billingsley (1979). Probability and Measure. New York, Toronto, London: John Wiley and Sons.
6. Olav Kallenberg; Foundations of Modern Probability, 2nd ed. Springer Series in Statistics. (2002). 650 pp. ISBN 0-387-95313-2
7. Henk Tijms (2004). Understanding Probability. Cambridge Univ. Press.
8. A lively introduction to probability theory for the beginner.
9. Olav Kallenberg; Probabilistic Symmetries and Invariance Principles. Springer -Verlag, New York (2005). 510 pp. ISBN 0-387-25115-4
10. Gut, Allan (2005). Probability: A Graduate Course. Springer-Verlag. ISBN 0-387-22833-0.

CHAPTER 4

Events in Probability Theory

Chapter Contents

In probability theory, an **event** is a set of outcomes of an experiment (a subset of the sample space) to which a probability is assigned. A single outcome may be an element of many different events, and different events in an experiment are usually not equally likely, since they may include very different groups of outcomes. An event defines a complementary event, namely the complementary set (the event *not* occurring), and together these define a Bernoulli trial: did the event occur or not?

Typically, when the sample space is finite, any subset of the sample space is an event (*i.e.* all elements of the power set of the sample space are defined as events). However, this approach does not work well in cases where the sample space is uncountably infinite, most notably when the outcome is a real number. So, when defining a probability space it is possible, and often necessary, to exclude certain subsets of the sample space from being events.

A SIMPLE EXAMPLE

If we assemble a deck of 52 playing cards with no jokers, and draw a single card from the deck, then the sample space is a 52-element set, as each card is a possible outcome. An event, however, is any subset of the sample space, including any singleton set (an elementary event), the empty set (an impossible event, with probability zero) and the sample space itself (a certain event, with probability one). Other events are proper subsets of the sample space that contain multiple elements. So, for example, potential events include:

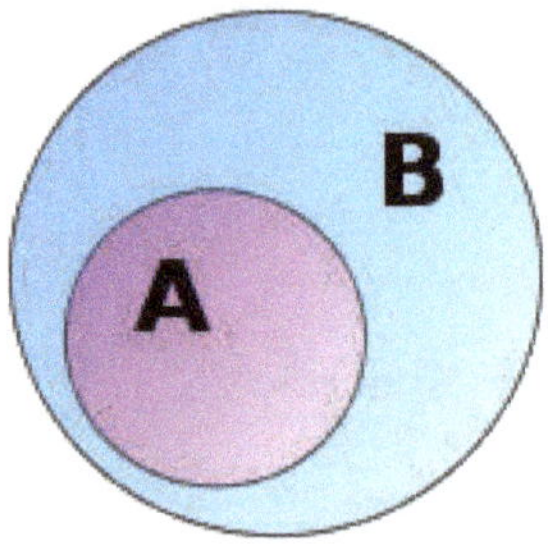

A Venn diagram of an event. *B* is the sample space and *A* is an event.

By the ratio of their areas, the probability of *A* is approximately 0.4.

- "Red and black at the same time without being a joker" (0 elements),
- "The 5 of Hearts" (1 element),
- "A King" (4 elements),
- "A Face card" (12 elements),
- "A Spade" (13 elements),
- "A Face card or a red suit" (32 elements),
- "A card" (52 elements).

Since all events are sets, they are usually written as sets (e.g. {1, 2, 3}), and represented graphically using Venn diagrams. In the situation where each outcome in the sample space Ω is equally likely, the probability P of an event *A* is the following formula:

$$\mathrm{P}(A) = \frac{|A|}{|\Omega|} \quad \left(\textbf{alternatively: } \mathrm{Pr}(A) = \frac{|A|}{|\Omega|}\right)$$

This rule can readily be applied to each of the example events above.

EVENTS IN PROBABILITY SPACES

Defining all subsets of the sample space as events works well when there are only finitely many outcomes, but gives rise to problems when the sample space is infinite. For many standard probability distributions, such as the normal distribution, the sample space is the set of real numbers or some subset of the real numbers. Attempts to define probabilities for all subsets of the real numbers run into difficulties when one considers 'badly behaved' sets, such as those that are nonmeasurable. Hence, it is necessary to restrict attention to a more limited family of subsets. For the standard tools of probability theory, such as joint and conditional probabilities, to work, it is necessary to use a σ-algebra, that is, a family closed under complementation and countable unions of its members. The most natural choice is the Borel measurable set derived from unions and intersections of intervals. However, the larger class of Lebesgue measurable sets proves more useful in practice.

In the general measure-theoretic description of probability spaces, an event may be defined as an element of a selected σ-algebra of subsets of the sample space. Under this definition, any subset of the sample space that is not an element of the σ-algebra is not an event, and does not have a probability. With a reasonable specification of the probability space, however, all *events of interest* are elements of the σ-algebra.

A NOTE ON NOTATION

Even though events are subsets of some sample space Ω, they are often written as propositional formulas involving random variables. For example, if X is a real-valued random variable defined on the sample space Ω, the event

$$\{\omega \in \Omega \mid u < X(\omega) \leq v\}$$

can be written more conveniently as, simply,

$$u < X \leq v.$$

This is especially common in formulas for a probability, such as

$$\Pr(u < X \leq v) = F(v) - F(u).$$

The set $u < X \leq v$ is an example of an inverse image under the mapping X because $\omega \in X^{-1}((u, v])$ if and only if $u < X(\omega) \leq v$.

ELEMENTARY EVENT

In probability theory, an **elementary event** (also called an **atomic event** or **simple event**) is an event which contains only a single outcome in the sample space. Using set theory terminology, an elementary event is a singleton. Elementary events and their corresponding outcomes are often written interchangeably for simplicity, as such an event corresponds to precisely one outcome.

The following are examples of elementary events:

- All sets {k}, where k ∈ N if objects are being counted and the sample space is S = {0, 1, 2, 3, ...} (the natural numbers).
- {HH}, {HT}, {TH} and {TT} if a coin is tossed twice. S = {HH, HT, TH, TT}. H stands for heads and T for tails.
- All sets {x}, where x is a real number. Here X is a random variable with a normal distribution and S = (−∞, +∞). This example shows that, because the probability of each elementary event is zero, the probabilities assigned to elementary events do not determine a continuous probability distribution.

PROBABILITY OF AN ELEMENTARY EVENT

Elementary events may occur with probabilities that are between zero and one (inclusively). In a discrete probability distribution whose sample space is finite, each elementary event is assigned a particular probability. In contrast, in a continuous distribution, individual elementary events must all have a probability of zero because there are infinitely many of them— then non-zero probabilities can only be assigned to non-elementary events.

Some "mixed" distributions contain both stretches of continuous elementary events and some discrete elementary events; the discrete elementary events in such distributions can be called **atoms** or **atomic events** and can have non-zero probabilities.

Under the measure-theoretic definition of a probability space, the probability of an elementary event need not even be defined. In particular, the set of events on which probability is defined may be some σ-algebra on S and not necessarily the full power set.

COMPLEMENTARY EVENT

In probability theory, the **complement** of any event A is the event [not A], i.e. the event that A does not occur. The event A and its complement [not A] are mutually exclusive and exhaustive. Generally, there is only one event B such that A and B are both mutually exclusive and exhaustive; that event is the complement of A. The complement of an event A is usually denoted as A', A^c or A. Given an event, the event and its complementary event define a Bernoulli trial: did the event occur or not?

For example, if a typical coin is tossed and one assumes that it cannot land on its edge, then it can either land showing "heads" or "tails." Because these two outcomes are mutually exclusive (i.e. the coin cannot simultaneously show both heads and tails) and collectively exhaustive (i.e. there are no other possible outcomes not represented between these two), they are therefore each other's complements. This means that [heads] is logically equivalent to [not tails], and [tails] is equivalent to [not heads].

Complement Rule

In a random experiment, the probabilities of all possible events (the sample space) must total to 1— that is, some outcome must occur on every trial. For two events to be complements, they must be collectively exhaustive, together filling the entire sample space. Therefore, the probability of an event's complement must be unity minus the probability of the event. That is, for an event A,

$$P(A^c) = 1 - P(A).$$

Equivalently, the probabilities of an event and its complement must always total to 1. This does not, however, mean that *any* two events whose probabilities total to 1 are each other›s complements; complementary events must also fulfill the condition of mutual exclusivity.

Example of the Utility of this Concept

Suppose one throws an ordinary six-sided die eight times. What is the probability that one sees a "1" at least once?

It may be tempting to say that

Pr(["1" on 1st trial] or ["1" on second trial] or ... or ["1" on 8th trial])

= Pr("1" on 1st trial) + Pr("1" on second trial) + ... + P("1" on 8th trial)

= 1/6 + 1/6 + ... + 1/6.

= 8/6 = 1.3333... (...and this is clearly wrong.)

That cannot be right because a probability cannot be more than 1. The technique is wrong because the eight events whose probabilities got added are not mutually exclusive.

One may resolve this overlap by the principle of inclusion-exclusion, or in this case one may instead more simply find the probability of the complementary event and subtract it from 1, thus:

Pr(at least one "1") = 1 − Pr(no "1"s)

= 1 − Pr([no "1" on 1st trial] and [no "1" on 2nd trial] and ... and [no "1" on 8th trial])

= 1 − Pr(no "1" on 1st trial) × Pr(no "1" on 2nd trial) × ... × Pr(no "1" on 8th trial)

= 1 −(5/6) × (5/6) × ... × (5/6)

= $1 - (5/6)^8$

= 0.7674...

NEGATION

In logic, **negation**, also called **logical complement**, is an operation that takes a proposition p to another proposition "not p", written $\neg p$, which is interpreted intuitively as being true when p is false and false when p is true. Negation is thus a unary (single-argument) logical connective. It may be applied as an operation on propositions, truth values, or semantic values more generally. In classical logic, negation is normally identified with the truth function that takes *truth* to *falsity* and vice versa. In intuitionistic logic, according to the Brouwer–Heyting–Kolmogorov interpretation, the negation of a proposition p is the proposition whose proofs are the refutations of p.

Definition

No agreement exists as to the possibility of defining negation, as to its logical status, function, and meaning, as to its field of applicability..., and as to the interpretation of the negative judgment, (F.H. Heinemann 1944).

Classical negation is an operation on one logical value, typically the value of a proposition, that produces a value of *true* when its operand is false and a value of *false* when its operand is true. So, if statement A is true, then $\neg A$ (pronounced «not A») would therefore be false; and conversely, if $\neg A$ is false, then A would be true.

The truth table of $\neg p$ is as follows:

Truth table of ¬p	
p	¬p
True	False
False	True

Classical negation can be defined in terms of other logical operations. For example, $\neg p$ can be defined as $p \rightarrow F$, where "→" is logical consequence and F is absolute falsehood. Conversely, one can define F as $p \,\&\, \neg p$ for any proposition p, where "&" is logical conjunction. The idea here is that any contradiction is false. While these ideas work in both classical and intuitionistic logic, they do not work in paraconsistent logic, where contradictions are not necessarily false. But in classical logic, we get a further identity: $p \rightarrow q$ can be defined as $\neg p \vee q$, where "∨" is logical disjunction: "not p, or q".

Algebraically, classical negation corresponds to complementation in a Boolean algebra, and intuitionistic negation to pseudocomplementation in a Heyting algebra. These algebras provide a semantics for classical and intuitionistic logic respectively.

Notation

The negation of a proposition p is notated in different ways in various contexts of discussion and fields of application. Among these variants are the following:

Notation	Vocalization
$\sim p$	Not p
$-p$	Not p
Np	En p
$\neg p$	Not p
p'	p prime, p complement
$\bar{p}$	p bar, Bar p
$!p$	Bang p Not p

In set theory \ is also used to indicate 'not member of': U \ A is the set of all members of U that are not members of A.

No matter how it is notated or symbolized, the negation $\neg p$ / $-p$ can be read as "it is not the case that p", "not that p", or usually more simply (though not grammatically) as "not p".

Properties

Double Negation

Within a system of classical logic, double negation, that is, the negation of the negation of a proposition p, is logically equivalent to p. Expressed in symbolic terms, $\neg\neg p \Leftrightarrow p$. In intuitionistic logic, a proposition implies its double negation but not conversely. This marks one important difference between classical and intuitionistic negation. Algebraically, classical negation is called an involution of period two.

However, in intuitionistic logic we do have the equivalence of $\neg\neg\neg p$ and $\neg p$. Moreover, in the propositional case, a sentence is classically provable if its double negation is intuitionistically provable. This result is known as Glivenko's theorem.

Distributivity

De Morgan's laws provide a way of distributing negation over disjunction and conjunction :

$\neg(a \vee b) \equiv (\neg a \wedge \neg b)$, and
$\neg(a \wedge b) \equiv (\neg a \vee \neg b)$.

Linearity

In Boolean algebra, a linear function is one such that:

If there exists $a_0, a_1, \ldots, a_n \in \{0,1\}$ such that $f(b_1, \ldots, b_n) = a_0 \oplus (a_1 \wedge b_1) \oplus \ldots \oplus (a_n \wedge b_n)$, for all $b_1, \ldots, b_n \in \{0,1\}$.

Another way to express this is that each variable always makes a difference in the truth-value of the operation or it never makes a difference. Negation is a linear logical operator.

Self dual

In Boolean algebra a self dual function is one such that:

$f(a_1, \ldots, a_n) = \sim f(\sim a_1, \ldots, \sim a_n)$ for all $a_1, \ldots, a_n \in \{0,1\}$. Negation is a self dual logical operator.

Rules of Inference

There are a number of equivalent ways to formulate rules for negation. One usual way to formulate classical negation in a natural deduction setting is to take as primitive rules of inference *negation introduction* (from a derivation of p to both q and $\neg q$, infer $\neg p$; this rule also being called *reductio ad absurdum*), *negation elimination* (from p and $\neg p$ infer q; this rule also being called *ex falso quodlibet*), and *double negation elimination* (from $\neg\neg p$ infer p). One obtains the rules for intuitionistic negation the same way but by excluding double negation elimination.

Negation introduction states that if an absurdity can be drawn as conclusion from p then p must not be the case (i.e. p is false (classically) or refutable (intuitionistically) or etc.). Negation elimination states that anything follows from an absurdity. Sometimes negation elimination is formulated using a primitive absurdity sign $\bot$. In this case the rule says that from p and $\neg p$ follows an absurdity. Together with double negation elimination one may infer our originally formulated rule, namely that anything follows from an absurdity.

Typically the intuitionistic negation $\neg p$ of p is defined as $p \to \bot$. Then negation introduction and elimination are just special cases of implication introduction (conditional proof) and elimination (modus ponens). In this case one must also add as a primitive rule *ex falso quodlibet.*

Programming

As in mathematics, negation is used in computer science to construct logical statements.

```
if (!(r == t))
{
    /*...statements executed when r does NOT equal t...*/
}
```

The "!" signifies logical NOT in B, C, and languages with a C-inspired syntax such as C++, Java, JavaScript, Perl, and PHP. "NOT" is the operator used in ALGOL 60, BASIC, and languages with an ALGOL- or BASIC-inspired syntax such as Pascal, Ada, Eiffel and Seed7. Some languages (C++, Perl, etc.) provide more than one operator for negation. A few languages like PL/I and Ratfor use ¬ for negation. Some modern computers and operating systems will display ¬ as ! on files encoded in ASCII. Most modern languages allow the above statement to be shortened from if (!(r == t)) to if (r != t), which allows sometimes, when the compiler/interpreter is not able to optimize it, faster programs.

In computer science there is also *bitwise negation.* This takes the value given and switches all the binary 1s to 0s and 0s to 1s. See bitwise operation. This is often used to create ones› complement or "~" in C or C++ and two›s complement (just simplified to "-" or the negative sign since this is equivalent to taking the arithmetic negative value of the number) as it basically creates the opposite (negative value equivalent) or mathematical complement of the value (where both values are added together they create a whole).

To get the absolute (positive equivalent) value of a given integer the following would work as the "-" changes it from negative to positive (it is negative because "x < 0" yields true)

unsigned int abs(int x)

```
{
    if (x < 0)
        return -x;
    else
        return x;
}
```

To demonstrate logical negation:

```
unsigned int abs(int x)
{
    if (!(x < 0))
        return x;
    else
        return -x;
}
```

Inverting the condition and reversing the outcomes produces code that is logically equivalent to the original code, i.e. will have identical results for any input (note that depending on the compiler used, the actual instructions performed by the computer may differ).

This convention occasionally surfaces in written speech, as computer-related slang for *not*. The phrase !voting, for example, means "not voting".

Kripke Semantics

In Kripke semantics where the semantic values of formulae are sets of possible worlds, negation can be taken to mean set-theoretic complementation. (See also possible world semantics.)

JOINT PROBABILITY DISTRIBUTION

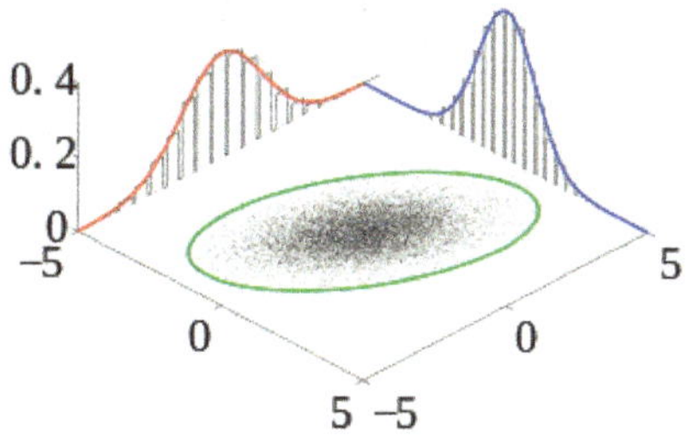

Many sample observations (black) are shown from a joint probability distribution. The marginal densities are shown as well.

In the study of probability, given at least two random variables *X*, *Y*, ..., that are defined on a probability space, the **joint probability distribution** for *X*, *Y*, ... is a probability distribution that gives the probability that each of *X*, *Y*, ... falls in any particular range or discrete set of values specified for that variable. In the case of only two random variables, this is called a **bivariate distribution**, but the concept generalizes to any number of random variables, giving a **multivariate distribution**.

The joint probability distribution can be expressed either in terms of a joint cumulative distribution function or in terms of a joint probability density function (in the case of continuous variables) or joint probability mass function (in the case of discrete variables). These in turn can be used to find two other types of distributions: the marginal distribution giving the probabilities for any one of the variables with no reference to any specific ranges of values for the other variables, and the conditional probability distribution giving the probabilities for any subset of the variables conditional on particular values of the remaining variables.

Examples

Coin Flips

Consider the flip of two fair coins; let *A* and *B* be discrete random variables associated with the outcomes first and second coin flips respectively. If a coin displays "heads" then associated random variable is 1, and is 0 otherwise. The joint probability mass function of *A* and *B* defines probabilities for each pair of outcomes. All possible outcomes are

$$(A = 0, B = 0), (A = 0, B = 1), (A = 1, B = 0), (A = 1, B = 1)$$

Since each outcome is equally likely the joint probability mass function becomes

$$P(A, B) = 1/4$$

when $A, B \in \{0, 1\}$. Since the coin flips are independent, the joint probability mass function is the product of the marginals:

$$P(A, B) = P(A)P(B).$$

In general, each coin flip is a Bernoulli trial and the sequence of flips follows a Bernoulli distribution.

Dice Rolls

Consider the roll of a fair die and let *A* = 1 if the number is even (i.e. 2, 4, or 6) and *A* = 0 otherwise. Furthermore, let *B* = 1 if the number is prime (i.e. 2, 3, or 5) and *B* = 0 otherwise.

	1	2	3	4	5	6
A	0	1	0	1	0	1
B	0	1	1	0	1	0

Then, the joint distribution of A and B, expressed as a probability mass function, is

$$\mathrm{P}(A=0,B=0)=P\{1\}=\frac{1}{6},\ \mathrm{P}(A=1,B=0)=P\{4,6\}=\frac{2}{6},$$

$$\mathrm{P}(A=0,B=1)=P\{3,5\}=\frac{2}{6},\ \mathrm{P}(A=1,B=1)=P\{2\}=\frac{1}{6}.$$

These probabilities necessarily sum to 1, since the probability of *some* combination of A and B occurring is 1.

Density Function or Mass Function

Discrete Case

The joint probability mass function of two discrete random variables X, Y is:

$$\mathrm{P}(X=x \text{ and } Y=y)=\mathrm{P}(Y=y \mid X=x)\cdot \mathrm{P}(X=x)=\mathrm{P}(X=x \mid Y=y)\cdot \mathrm{P}(Y=y),$$

where $\mathrm{P}(Y=y \mid X=x)$ is the probability of $Y=y$ given that $X=x$.

The generalization of the preceding two variable case is the joint probability distribution of n discrete random variables $X_1, X_2, \ldots, X_n$ which is:

$$\begin{aligned}\mathrm{P}(X_1=x_1,\ldots,X_n=x_n)=&\ \mathrm{P}(X_1=x_1)\times \mathrm{P}(X_2=x_2 \mid X_1=x_1)\\ &\times \mathrm{P}(X_3=x_3 \mid X_1=x_1,X_2=x_2)\\ &\ldots\\ &\times P(X_n=x_n \mid X_1=x_1,X_2=x_2,\ldots,X_{n-1}=x_{n-1})\end{aligned}$$

This identity is known as the chain rule of probability.

Since these are probabilities, we have in the two-variable case

$$\sum_i\sum_j \mathrm{P}(X=x_i \text{ and } Y=y_j)=1,$$

which generalizes for n discrete random variables $X_1, X_2, \ldots, X_n$ to

$$\sum_i\sum_j\ldots\sum_k \mathrm{P}(X_1=x_{1i},X_2=x_{2j},\ldots,X_n=x_{nk})=1.$$

Continuous Case

The **joint probability density function** $f_{X,Y}(x, y)$ for continuous random variables is equal to:

$$f_{X,Y}(x,y)=f_{Y|X}(y \mid x)f_X(x)=f_{X|Y}(x \mid y)f_Y(y)$$

...where $f_{Y|X}(y|x)$ and $f_{X|Y}(x|y)$ give the conditional distributions of Y given $X = x$ and of X given $Y = y$ respectively, and $f_X(x)$ and $f_Y(y)$ give the marginal distributions for X and Y respectively.

Again, since these are probability distributions, one has

$$\int_x \int_y f_{X,Y}(x,y)\, dy\, dx = 1.$$

Mixed Case

The "mixed joint density" may be defined where one random variable X is continuous and the other random variable Y is discrete, or vice versa, as:

$$f_{X,Y}(x,y) = f_{X|Y}(x \mid y)\mathrm{P}(Y=y) = \mathrm{P}(Y=y \mid X=x) f_X(x)$$

One example of a situation in which one may wish to find the cumulative distribution of one random variable which is continuous and another random variable which is discrete arises when one wishes to use a logistic regression in predicting the probability of a binary outcome Y conditional on the value of a continuously distributed outcome X. One *must* use the "mixed" joint density when finding the cumulative distribution of this binary outcome because the input variables (X, Y) were initially defined in such a way that one could not collectively assign it either a probability density function or a probability mass function. Formally, $f_{X,Y}(x, y)$ is the probability density function of (X, Y) with respect to the product measure on the respective supports of X and Y. Either of these two decompositions can then be used to recover the joint cumulative distribution function:

$$F_{X,Y}(x,y) = \sum_{t \le y} \int_{s=-\infty}^{x} f_{X,Y}(s,t)\, ds$$

The definition generalizes to a mixture of arbitrary numbers of discrete and continuous random variables.

Additional Properties

Joint distribution for Independent Variables

Two discrete random variables X and Y are independent if the joint probability mass function satisfies

$$P(X = x \text{ and } Y = y) = P(X = x) \cdot P(Y = y)$$

for all x and y.

Similarly, two absolutely continuous random variables are independent if

$$f_{X,Y}(x,y) = f_X(x) \cdot f_Y(y)$$

for all x and y. This means that acquiring any information about the value of one or more of the random variables leads to a conditional distribution of any other variable that is identical to its unconditional (marginal) distribution; thus no variable provides any information about any other variable.

Joint Distribution for Conditionally Dependent Variables

If a subset A of the variables $X_1, \cdots, X_n$ is conditionally dependent given another subset B of these variables, then the joint distribution $\mathrm{P}(X_1, \ldots, X_n)$ is equal to $P(B) \cdot P(A \mid B)$. Therefore, it can be efficiently represented by the lower-dimensional probability distributions $P(B)$ and $P(A \mid B)$. Such conditional independence relations can be represented with a Bayesian network or copula functions.

Cumulative Distribution

The joint probability distribution for a pair of random variables can be expressed in terms of their cumulative distribution function $F(x, y) = P(X \leq x, Y \leq y)$.

Important Named Distributions

Named joint distributions that arise frequently in statistics include the multivariate normal distribution, the multivariate stable distribution, the multinomial distribution, the negative multinomial distribution, the multivariate hypergeometric distribution, and the elliptical distribution.

BINOMIAL DISTRIBUTION

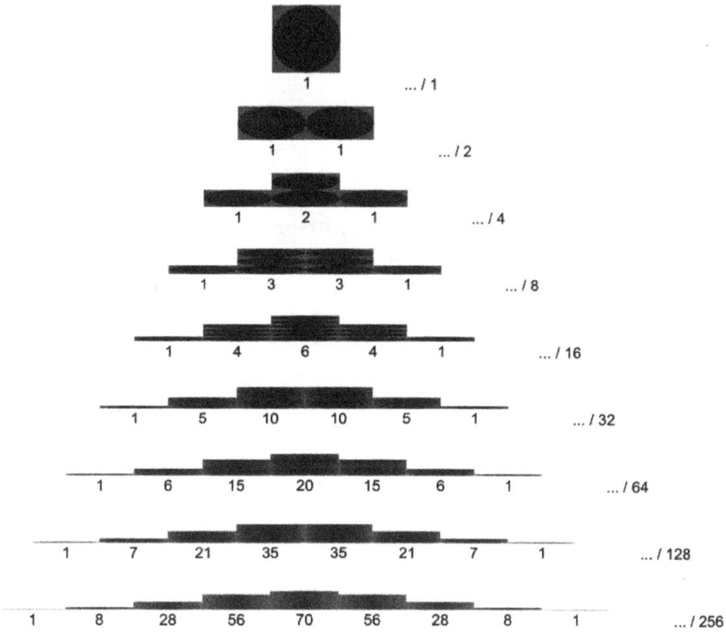

Binomial distribution for $p = 0.5$

with *n* and *k* as in Pascal›s triangle

The probability that a ball in a Galton box with 8 layers ($n = 8$) ends up in the central bin ($k = 4$) is $(k = 4)$ is $70/256$.

In probability theory and statistics, the **binomial distribution** with parameters n and p is the discrete probability distribution of the number of successes in a sequence of n independent yes/no experiments, each of which yields success with probability p. A success/failure experiment is also called a Bernoulli experiment or Bernoulli trial; when $n = 1$, the binomial distribution is a Bernoulli distribution. The binomial distribution is the basis for the popular binomial test of statistical significance.

The binomial distribution is frequently used to model the number of successes in a sample of size n drawn with replacement from a population of size N. If the sampling is carried out without replacement, the draws are not independent and so the resulting distribution is a hypergeometric distribution, not a binomial one. However, for N much larger than n, the binomial distribution remains a good approximation, and is widely used.

Specification

Probability Mass Function

In general, if the random variable X follows the binomial distribution with parameters $n \in \mathbb{N}$ and $p \in [0,1]$, we write $X \sim B(n, p)$. The probability of getting exactly k successes in n trials is given by the probability mass function:

$$f(k; n, p) = \Pr(X = k) = \binom{n}{k} p^k (1-p)^{n-k}$$

for $k = 0, 1, 2, \ldots, n$, where

$$\binom{n}{k} = \frac{n!}{k!(n-k)!}$$

is the binomial coefficient, hence the name of the distribution. The formula can be understood as follows. k successes occur with probability p^k and $n - k$ failures occur with probability $(1 - p)^{n-k}$. However, the k successes can occur anywhere among the n trials, and there are $\binom{n}{k}$ different ways of distributing k successes in a sequence of n trials.

In creating reference tables for binomial distribution probability, usually the table is filled in up to $n/2$ values. This is because for $k > n/2$, the probability can be calculated by its complement as

$$f(k, n, p) = f(n - k, n, 1 - p).$$

The probability mass function satisfies the following recurrence relation, for every n,p:

$$\left\{\begin{array}{l} p(n-k)f(k,n,p) = (k+1)(1-p)f(k+1,n,p), \\ f(0,n,p) = (1-p)^n \end{array}\right\}$$

Looking at the expression $f(k, n, p)$ as a function of k, there is a k value that maximizes it. This k value can be found by calculating

$$\frac{f(k+1,n,p)}{f(k,n,p)} = \frac{(n-k)p}{(k+1)(1-p)}$$

and comparing it to 1. There is always an integer M that satisfies

$$(n+1)p - 1 \leq M < (n+1)p.$$

$f(k, n, p)$ is monotone increasing for $k < M$ and monotone decreasing for $k > M$, with the exception of the case where $(n + 1)p$ is an integer. In this case, there are two values for which f is maximal: $(n + 1)p$ and $(n + 1)p - 1$. M is the *most probable* (*most likely*) outcome of the Bernoulli trials and is called the mode. Note that the probability of it occurring can be fairly small.

Cumulative Distribution Function

The cumulative distribution function can be expressed as:

$$F(k;n,p) = \Pr(X \leq k) = \sum_{i=0}^{\lfloor k \rfloor} \binom{n}{i} p^i (1-p)^{n-i}$$

where $\lfloor k \rfloor$ is the "floor" under k, i.e. the greatest integer less than or equal to k.

It can also be represented in terms of the regularized incomplete beta function, as follows:

$$\begin{aligned} F(k;n,p) &= \Pr(X \leq k) \\ &= I_{1-p}(n-k, k+1) \\ &= (n-k)\binom{n}{k} \int_0^{1-p} t^{n-k-1}(1-t)^k \, dt. \end{aligned}$$

Some closed-form bounds for the cumulative distribution function are given below.

Example

Suppose a biased coin comes up heads with probability 0.3 when tossed. What is the probability of achieving 0, 1,..., 6 heads after six tosses?

$$\Pr(0\text{ heads}) = f(0) = \Pr(X=0) = \binom{6}{0}0.3^0(1-0.3)^{6-0} \approx 0.1176$$

$$\Pr(1\text{ heads}) = f(1) = \Pr(X=1) = \binom{6}{1}0.3^1(1-0.3)^{6-1} \approx 0.3025$$

$$\Pr(2\text{ heads}) = f(2) = \Pr(X=2) = \binom{6}{2}0.3^2(1-0.3)^{6-2} \approx 0.3241$$

$$\Pr(3\text{ heads}) = f(3) = \Pr(X=3) = \binom{6}{3}0.3^3(1-0.3)^{6-3} \approx 0.1852$$

$$\Pr(4\text{ heads}) = f(4) = \Pr(X=4) = \binom{6}{4}0.3^4(1-0.3)^{6-4} \approx 0.0595$$

$$\Pr(5\text{ heads}) = f(5) = \Pr(X=5) = \binom{6}{5}0.3^5(1-0.3)^{6-5} \approx 0.0102$$

$$\Pr(6\text{ heads}) = f(6) = \Pr(X=6) = \binom{6}{6}0.3^6(1-0.3)^{6-6} \approx 0.0007$$

Mean

If $X \sim B(n, p)$, that is, X is a binomially distributed random variable, n being the total number of experiments and p the probability of each experiment yielding a successful result, then the expected value of X is:

$$\mathrm{E}[X] = np.$$

For example, if $n = 100$, and $p = 1/4$, then the average number of successful results will be 25.

Proof: We calculate the mean, μ, directly calculated from its definition

$$\mu = \sum_{i=1}^{n} x_i p_i,$$

and the binomial theorem:

$$\begin{aligned}
\mu &= \sum_{k=0}^{n} k\binom{n}{k}p^k(1-p)^{n-k} \\
&= np\sum_{k=0}^{n} k\frac{(n-1)!}{(n-k)!k!}p^{k-1}(1-p)^{(n-1)-(k-1)} \\
&= np\sum_{k=1}^{n} \frac{(n-1)!}{((n-1)-(k-1))!(k-1)!}p^{k-1}(1-p)^{(n-1)-(k-1)} \\
&= np\sum_{k=1}^{n} \binom{n-1}{k-1}p^{k-1}(1-p)^{(n-1)-(k-1)} \\
&= np\sum_{\ell=0}^{n-1} \binom{n-1}{\ell}p^{\ell}(1-p)^{(n-1)-\ell} && \text{with } \ell := k-1 \\
&= np\sum_{\ell=0}^{m} \binom{m}{\ell}p^{\ell}(1-p)^{m-\ell} && \text{with } m := n-1 \\
&= np(p+(1-p))^m \\
&= np
\end{aligned}$$

It is also possible to deduce the mean from the equation $X = X_1 + \cdots + X_n$ whereby all X_i

are Bernoulli distributed random variables with $E[X_i] = p$. We get

$$E[X] = E[X_1 + \cdots + X_n] = E[X_1] + \cdots + E[X_n] = \underbrace{p + \cdots + p}_{n \text{ times}} = np$$

Variance

The variance is:

$$\operatorname{Var}(X) = np(1-p).$$

Proof: Let $X = X_1 + \cdots + X_n$ where all X_i are independently Bernoulli distributed random variables. We get because of $\operatorname{Var}(X_i) = p(1-p)$:

$$\operatorname{Var}(X) = \operatorname{Var}(X_1 + \cdots + X_n) = \operatorname{Var}(X_1) + \cdots + \operatorname{Var}(X_n) = n\operatorname{Var}(X_1) = np(1-p).$$

Mode

Usually the mode of a binomial $B(n,p)$ distribution is equal to $\lfloor (n+1)p \rfloor$, where $\lfloor \cdot \rfloor$ is the floor function. However, when $(n + 1)p$ is an integer and p is neither 0 nor 1, then the distribution has two modes: $(n + 1)p$ and $(n + 1)p - 1$. When p is equal to 0 or 1, the mode will be 0 and n correspondingly. These cases can be summarized as follows:

$$\text{mode} = \begin{cases} \lfloor (n+1)\,p \rfloor & \text{if } (n+1)p \text{ is 0 or a noninteger}, \\ (n+1)\,p \ \text{ and } \ (n+1)\,p-1 & \text{if } (n+1)p \in \{1,\dots,n\}, \\ n & \text{if } (n+1)p = n+1. \end{cases}$$

Proof: Let

$$f(k) = \binom{n}{k} p^k q^{n-k}.$$

For $p = 0$ only $f(0)$ has a nonzero value with $f(0) = 1$. For $p = 1$ we find $f(n) = 1$ and $f(k) = 0$ for $k \neq n$. This proves that the mode is 0 for $p = 0$ and n for $p = 1$.

Let $0 < p < 1$. We find

$$\frac{f(k+1)}{f(k)} = \frac{(n-k)p}{(k+1)(1-p)}.$$

From this follows

$$k > (n+1)p - 1 \Rightarrow f(k+1) < f(k)$$
$$k = (n+1)p - 1 \Rightarrow f(k+1) = f(k)$$
$$k < (n+1)p - 1 \Rightarrow f(k+1) > f(k)$$

So when $(n+1)p-1$ is an integer, then $(n+1)p-1$ and $(n+1)p$ is a mode. In the case that $(n+1)p-1 \notin \mathbb{Z}$, then only $\lfloor (n+1)p-1 \rfloor + 1 = \lfloor (n+1)p \rfloor$ is a mode.[4]

Median

In general, there is no single formula to find the median for a binomial distribution, and it may even be non-unique. However several special results have been established:

- If np is an integer, then the mean, median, and mode coincide and equal np.
- Any median m must lie within the interval $\lfloor np \rfloor \leq m \leq \lceil np \rceil$.
- A median m cannot lie too far away from the mean: $|m - np| \leq \min\{\ln 2, \max\{p, 1 - p\}\}$.
- The median is unique and equal to m = round(np) in cases when either $p \leq 1 - \ln 2$ or $p \geq \ln 2$ or $|m - np| \leq \min\{p, 1 - p\}$ (except for the case when $p = ½$ and n is odd).
- When $p = 1/2$ and n is odd, any number m in the interval $½(n - 1) \leq m \leq ½(n + 1)$ is a median of the binomial distribution. If $p = 1/2$ and n is even, then $m = n/2$ is the unique median.

Covariance between two binomials

If two binomially distributed random variables X and Y are observed together, estimating their covariance can be useful. Using the definition of covariance, in the case $n = 1$ (thus being Bernoulli trials) we have

$$\text{Cov}(X, Y) = \text{E}(XY) - \mu_X \mu_Y.$$

The first term is non-zero only when both X and Y are one, and μ_X and μ_Y are equal to the two probabilities. Defining p_B as the probability of both happening at the same time, this gives

$$\text{Cov}(X, Y) = p_B - p_X p_Y,$$

and for n independent pairwise trials

$$\text{Cov}(X, Y)_n = n(p_B - p_X p_Y)$$

If X and Y are the same variable, this reduces to the variance formula given above.

RELATED DISTRIBUTIONS

Sums of Binomials

If $X \sim$ B(n, p) and $Y \sim$ B(m, p) are independent binomial variables with the same probability p, then $X + Y$ is again a binomial variable; its distribution is $Z=X+Y \sim$ B($n+m$, p):

$$\begin{aligned} P(Z=k) &= \sum_{i=0}^{k}\left[\binom{n}{i}p^i(1-p)^{n-i}\right]\left[\binom{m}{k-i}p^{k-i}(1-p)^{m-k+i}\right] \\ &= \binom{n+m}{k}p^k(1-p)^{n+m-k} \end{aligned}$$

However, if X and Y do not have the same probability p, then the variance of the sum will be smaller than the variance of a binomial variable distributed as $B(n+m,\bar{p})$.

I

If $X \sim$ B(n, p) and, conditional on X, $Y \sim$ B(X, q), then Y is a simple binomial variable with distribution

$$Y \sim B(n, pq).$$

For example, imagine throwing n balls to a basket U_X and taking the balls that hit and throwing them to another basket U_Y. If p is the probability to hit U_X then $X \sim$ B(n, p) is the number of balls that hit U_X. If q is the probability to hit U_Y then the number of balls that hit U_Y is $Y \sim$ B(X, q) and therefore $Y \sim$ B(n, pq).

Bernoulli Distribution

The Bernoulli distribution is a special case of the binomial distribution, where $n = 1$. Symbolically, $X \sim$ B(1, p) has the same meaning as $X \sim$ B(p). Conversely, any binomial distribution, B(n, p), is the distribution of the sum of n Bernoulli trials, B(p), each with the same probability p.

Poisson Binomial Distribution

The binomial distribution is a special case of the Poisson binomial distribution, or general binomial distribution, which is the distribution of a sum of n independent non-identical Bernoulli trials B(p_i).

Normal Approximation

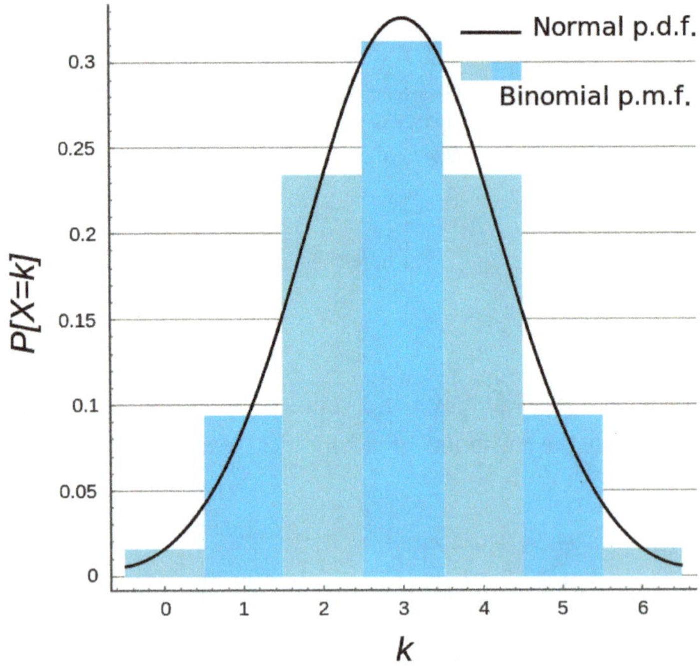

Binomial probability mass function and normal probability density function approximation for $n = 6$ and $p = 0.5$

If n is large enough, then the skew of the distribution is not too great. In this case a reasonable approximation to $B(n, p)$ is given by the normal distribution

$$\mathcal{N}(np, np(1-p)),$$

and this basic approximation can be improved in a simple way by using a suitable continuity correction. The basic approximation generally improves as n increases (at least 20) and is better when p is not near to 0 or 1. Various rules of thumb may be used to decide whether n is large enough, and p is far enough from the extremes of zero or one:

- One rule is that for $n > 5$ the normal approximation is adequate if the absolute value of the skewness is strictly less than 1/3; that is, if

$$\frac{|1-2p|}{\sqrt{np(1-p)}} = \frac{1}{\sqrt{n}}\left|\sqrt{\frac{1-p}{p}} - \sqrt{\frac{p}{1-p}}\right| < \frac{1}{3}.$$

- A stronger rule states that the normal approximation is appropriate only if everything within 3 standard deviations of its mean is within the range of possible values; that is, only if

$$\mu \pm 3\sigma = np \pm 3\sqrt{np(1-p)} \in (0, n).$$

This 3-standard-deviation rule is equivalent to the following conditions, which also imply the first rule above.

$$n > 9\frac{1-p}{p} \quad \text{and} \quad n > 9\frac{p}{1-p}.$$

[Proof]

- Another commonly used rule is that both values np and $n(1-p)$ must be greater than 5. However, the specific number varies from source to source, and depends on how good an approximation one wants. Some sources request 9 instead of 5, which gives the same result stated in the previous rule.

The rule $np \pm 3\sqrt{np(1-p)} \in (0, n)$ is totally equivalent to request that

$$np - 3\sqrt{np(1-p)} > 0 \quad \text{and} \quad np + 3\sqrt{np(1-p)} < n.$$

Moving terms around yields:

$$np > 3\sqrt{np(1-p)} \quad \text{and} \quad n(1-p) > 3\sqrt{np(1-p)}.$$

Since $0 < p < 1$, we can apply the square power and divide by the respective factors np^2 and $n(1-p)^2$, to obtain the desired conditions:

$$n > 9\frac{1-p}{p} \quad \text{and} \quad n > 9\frac{p}{1-p}.$$

Notice that these conditions automatically imply that $n > 9$. On the other hand, apply again the square root and divide by 3,

$$\frac{\sqrt{n}}{3} > \sqrt{\frac{1-p}{p}} > 0 \quad \text{and} \quad \frac{\sqrt{n}}{3} > \sqrt{\frac{p}{1-p}} > 0.$$

Subtracting the second set of inequalities from the first one yields:

$$\frac{\sqrt{n}}{3} > \sqrt{\frac{1-p}{p}} - \sqrt{\frac{p}{1-p}} > -\frac{\sqrt{n}}{3};$$

and so, the desired first rule is satisfied,

$$\left|\sqrt{\frac{1-p}{p}} - \sqrt{\frac{p}{1-p}}\right| < \frac{\sqrt{n}}{3}.$$

[Proof]

The following is an example of applying a continuity correction. Suppose one wishes to calculate $\Pr(X \leq 8)$ for a binomial random variable X. If Y has a distribution given by the normal approximation, then $\Pr(X \leq 8)$ is approximated by $\Pr(Y \leq 8.5)$. The addition of 0.5 is the continuity correction; the uncorrected normal approximation gives considerably less accurate results.

Assume that both values np and $n(1-p)$ are greater than 9. Since $0 < p < 1$, we easily have that

$$np \geq 9 > 9(1-p) \quad \text{and} \quad n(1-p) \geq 9 > 9p.$$

We only have to divide now by the respective factors p and $1-p$, to deduce the alternative form of the 3-standard-deviation rule:

$$n > 9\frac{1-p}{p} \quad \text{and} \quad n > 9\frac{p}{1-p}.$$

This approximation, known as de Moivre–Laplace theorem, is a huge time-saver when undertaking calculations by hand (exact calculations with large n are very onerous); historically, it was the first use of the normal distribution, introduced in Abraham de Moivre›s book *The Doctrine of Chances* in 1738. Nowadays, it can be seen as a consequence of the central limit theorem since $B(n, p)$ is a sum of n independent, identically distributed Bernoulli variables with parameter p. This fact is the basis of a hypothesis test, a "proportion z-test", for the value of p using x/n, the sample proportion and estimator of p, in a common test statistic.

For example, suppose one randomly samples n people out of a large population and ask them whether they agree with a certain statement. The proportion of people who agree will of course depend on the sample. If groups of n people were sampled repeatedly and truly randomly, the proportions would follow an approximate normal distribution with mean equal to the true proportion p of agreement in the population and with standard deviation $\sigma = \sqrt{\frac{p(1-p)}{n}}$.

Poisson Approximation

The binomial distribution converges towards the Poisson distribution as the number of trials goes to infinity while the product np remains fixed or at least p tends to zero. Therefore, the Poisson distribution with parameter $\lambda = np$ can be used as an approximation to $B(n, p)$ of the binomial distribution if n is sufficiently large and p is sufficiently small. According to two rules of thumb, this approximation is good if $n \geq 20$ and $p \leq 0.05$, or if $n \geq 100$ and $np \leq 10$.

Concerning the accuracy of Poisson approximation, see Novak, ch. 4, and references therein.

Limiting Distributions

- *Poisson limit theorem*: As n approaches ∞ and p approaches 0, then the Binomial(n, p) distribution approaches the Poisson distribution with expected value $\lambda = np$.
- *de Moivre–Laplace theorem*: As n approaches ∞ while p remains fixed, the distribution of

$$\frac{X-np}{\sqrt{np(1-p)}}$$

approaches the normal distribution with expected value 0 and variance 1. This result is sometimes loosely stated by saying that the distribution of X is asymptotically normal with expected value np and variance $np(1 - p)$. This result is a specific case of the central limit theorem.

Beta Distribution

Beta distributions provide a family of prior probability distributions for binomial distributions in Bayesian inference:

$$P(p;\alpha,\beta) = \frac{p^{\alpha-1}(1-p)^{\beta-1}}{B(\alpha,\beta)}.$$

CONFIDENCE INTERVALS

Even for quite large values of n, the actual distribution of the mean is significantly nonnormal.Because of this problem several methods to estimate confidence intervals have been proposed.

Let n_1 be the number of successes out of n, the total number of trials, and let

$$\hat{p} = \frac{n_1}{n}$$

be the proportion of successes. Let $z_{\alpha/2}$ be the $100(1 - \alpha/2)$th percentile of the standard normal distribution.

- Wald method

$$\hat{p} \pm z_{\frac{\alpha}{2}}\sqrt{\frac{\hat{p}(1-\hat{p})}{n}}.$$

A continuity correction of $0.5/n$ may be added.

- Agresti-Coull method

$$\tilde{p} \pm z_{\frac{\alpha}{2}}\sqrt{\frac{\tilde{p}(1-\tilde{p})}{n+z^2_{\frac{\alpha}{2}}}}.$$

Here the estimate of p is modified to

$$\tilde{p} = \frac{n_1 + \frac{1}{2}z^2_{\frac{\alpha}{2}}}{n+z^2_{\frac{\alpha}{2}}}$$

- ArcSine method

$$\sin^2\left(\arcsin(\sqrt{\hat{p}}) \pm \frac{z}{2\sqrt{n}}\right)$$

- Wilson (score) method

$$\frac{\hat{p} + \frac{1}{2n} z^2_{1-\frac{\alpha}{2}} \pm \frac{1}{2n} z_{1-\frac{\alpha}{2}} \sqrt{4n\hat{p}(1-\hat{p}) + z^2_{1-\frac{\alpha}{2}}}}{1 + \frac{1}{n} z^2_{1-\frac{\alpha}{2}}}.$$

The exact (Clopper-Pearson) method is the most conservative. The Wald method although commonly recommended in the text books is the most biased.

Generating Binomial Random Variates

Methods for random number generation where the marginal distribution is a binomial distribution are well-established.

One way to generate random samples from a binomial distribution is to use an inversion algorithm. To do so, one must calculate the probability that P(X=k) for all values *k* from 0 through *n*. (These probabilities should sum to a value close to one, in order to encompass the entire sample space.) Then by using a pseudorandom number generator to generate samples uniformly between 0 and 1, one can transform the calculated samples U[0,1] into discrete numbers by using the probabilities calculated in step one.

Tail Bounds

For $k \leq np$, upper bounds for the lower tail of the distribution function can be derived. Recall that $F(k;n,p) = \Pr(X \leq k)$, the probability that there are at most *k* successes.

Hoeffding›s inequality yields the bound

$$F(k;n,p) \leq \exp\left(-2\frac{(np-k)^2}{n}\right)$$

and Chernoff›s inequality can be used to derive the bound

$$F(k;n,p) \leq \exp\left(-\frac{1}{2p}\frac{(np-k)^2}{n}\right)$$

Moreover, these bounds are reasonably tight when $p = 1/2$, since the following expression holds for all $k \geq 3n/8$

$$F(k;n,\tfrac{1}{2}) \leq \frac{14}{15}\exp\left(-\frac{16(\frac{n}{2}-k)^2}{n}\right)$$

However, the bounds do not work well for extreme values of *p*. In particular, as $p \to 1$, value $F(k;n,p)$ goes to zero (for fixed *k*, *n* with $k<n$) while the upper bound above goes to a positive constant. In this case a better bound is given by

$$F(k;n,p) \leq \exp\left(-nD\left(\frac{k}{n}\|p\right)\right) \qquad \text{if } 0 < \frac{k}{n} < p$$

where $D(a\|p)$ is the relative entropy between an *a*-coin and a *p*-coin (i.e. between the Bernoulli(a) and Bernoulli(p) distribution):

$$F(k;n,p) \leq \exp\left(-nD\left(\frac{k}{n}||p\right)\right) \qquad \text{if } 0 < \frac{k}{n} < p$$

Asymptotically, this bound is reasonably tight; see for details. An equivalent formulation of the bound is

$$D(a||p) = (a)\log\frac{a}{p} + (1-a)\log\frac{1-a}{1-p}.$$

Both these bounds are derived directly from the Chernoff bound. It can also be shown that,

$$\Pr(X \geq k) = F(n-k;n,1-p) \leq \exp\left(-nD\left(\frac{k}{n}||p\right)\right) \qquad \text{if } p < \frac{k}{n} < 1.$$

This is proved using the method of types (see for example chapter 12 of Elements of Information Theory by Cover and Thomas).

We can also change the $(n+1)^2$ in the denominator to $\sqrt{2n}$, by approximating the binomial coefficient with Stirlings formula.

REFERENCES

1. Wadsworth, G. P. (1960). Introduction to Probability and Random Variables. New York: McGraw-Hill. p. 52.
2. Hamilton Institute. “The Binomial Distribution” October 20, 2010.
3. See Proof Wiki
4. See also the answer to the question “finding mode in Binomial distribution”
5. Neumann, P. (1966). “Über den Median der Binomial- and Poissonverteilung”. Wissenschaftliche Zeitschrift der Technischen Universität Dresden (in German). 19: 29–33.
6. Lord, Nick. (July 2010). “Binomial averages when the mean is an integer”, The Mathematical Gazette 94, 331-332.
7. Kaas, R.; Buhrman, J.M. (1980). “Mean, Median and Mode in Binomial Distributions”. Statistica Neerlandica. 34 (1): 13–18. doi:10.1111/j.1467-9574.1980.tb00681.x.
8. Hamza, K. (1995). “The smallest uniform upper bound on the distance between the mean and the median of the binomial and Poisson distributions”. Statistics & Probability Letters. 23: 21–25. doi:10.1016/0167-7152(94)00090-U.
9. Wang, Y. H. (1993). “On the number of successes in independent trials” (PDF). Statistica Sinica. 3 (2): 295–312.
10. Box, Hunter and Hunter (1978). Statistics for experimenters. Wiley. p. 130.
11. NIST/SEMATECH, “7.2.4. Does the proportion of defectives meet requirements?” e-Handbook of Statistical Methods.
12. NIST/SEMATECH, “6.3.3.1. Counts Control Charts”, e-Handbook of Statistical Methods.

13. Novak S.Y. (2011) Extreme value methods with applications to finance. London: CRC/ Chapman & Hall/Taylor & Francis. ISBN 9781-43983-5746.

14. MacKay, David (2003). Information Theory, Inference and Learning Algorithms. Cambridge University Press; First Edition. ISBN 978-0521642989.

15. Brown, Lawrence D.; Cai, T. Tony; DasGupta, Anirban (2001), "Interval Estimation for a Binomial Proportion", Statistical Science, 16 (2): 101–133, doi:10.1214/ss/1009213286, retrieved 2015-01-05

16. Agresti, Alan; Coull, Brent A. (May 1998), "Approximate is better than ‹exact› for interval estimation of binomial proportions" (PDF), The American Statistician, 52 (2): 119–126, doi:10.2307/2685469, retrieved 2015-01-05

17. Pires MA Confidence intervals for a binomial proportion: comparison of methods and software evaluation.

18. Wilson, Edwin B. (June 1927), "Probable inference, the law of succession, and statistical inference" (PDF), J. American Statistical Association, 22 (158): 209–212, doi:10.2307/2276774, retrieved 2015-01-05

19. Devroye, Luc (1986) Non-Uniform Random Variate Generation, New York: Springer-Verlag. (See especially Chapter X, Discrete Univariate Distributions)

20. Kachitvichyanukul, V.; Schmeiser, B. W. (1988). "Binomial random variate generation". Communications of the ACM. 31 (2): 216–222. doi:10.1145/42372.42381.

21. Matoušek, J, Vondrak, J: The Probabilistic Method (lecture notes) .

22. R. Arratia and L. Gordon: Tutorial on large deviations for the binomial distribution, Bulletin of Mathematical Biology 51(1) (1989), 125–131 .

23. Theorem 11.1.3 in Cover, T.; Thomas, J. (2006). Elements of Information Theory (2nd ed.). Wiley. p. 350.

24. http://math.stackexchange.com/questions/1548940/sharper-lower-bounds-for-binomial-chernoff-tails/1564088#1564088

25. Mandelbrot, B. B., Fisher, A. J., & Calvet, L. E. (1997). A multifractal model of asset returns. 3.2 The Binomial Measure is the Simplest Example of a Mu*ltifractal*

CHAPTER 5

Independence Probability Theory

Chapter Contents

- Introduction
- Definition
- Properties
- Examples
- Conditional Independence
- Graphoid
- Types of Graphoids

INTRODUCTION

In probability theory, two events are independent, statistically independent, or stochastically independent if the occurrence of one does not affect the probability of the other. Similarly, two random variables are independent if the realization of one does not affect the probability distribution of the other.

The concept of independence extends to dealing with collections of more than two events or random variables, in which case the events are pairwise independent if each pair are independent of each other, and the events are mutually independent if each event is independent of each other combination of events.

DEFINITION

For Events

Two Events

Two events A and B are **independent** (often written as $A \perp B$ or $A \perp\!\!\!\perp B$) if their joint probability equals the product of their probabilities:

$$\mathrm{P}(A \cap B) = \mathrm{P}(A)\mathrm{P}(B).$$

Why this defines independence is made clear by rewriting with conditional probabilities:

$$\mathrm{P}(A \cap B) = \mathrm{P}(A)\mathrm{P}(B) \Leftrightarrow \mathrm{P}(A) = \frac{\mathrm{P}(A)\mathrm{P}(B)}{\mathrm{P}(B)} = \frac{\mathrm{P}(A \cap B)}{\mathrm{P}(B)} = \mathrm{P}(A \mid B).$$

and similarly

$$\mathrm{P}(A \cap B) = \mathrm{P}(A)\mathrm{P}(B) \Leftrightarrow \mathrm{P}(B) = \mathrm{P}(B \mid A).$$

Thus, the occurrence of B does not affect the probability of A, and vice versa. Although the derived expressions may seem more intuitive, they are not the preferred definition, as the conditional probabilities may be undefined if $P(A)$ or $P(B)$ are 0. Furthermore, the preferred definition makes clear by symmetry that when A is independent of B, B is also independent of A.

More than Two Events

A finite set of events $\{A_i\}$ is **pairwise independent** if and only if every pair of events is independent—that is, if and only if for all distinct pairs of indices m, k,

$$\mathrm{P}(A_m \cap A_k) = \mathrm{P}(A_m)\mathrm{P}(A_k).$$

A finite set of events is **mutually independent** if and only if every event is independent of any intersection of the other events—that is, if and only if for every n-element subset $\{A_i\}$,

$$\mathrm{P}\left(\bigcap_{i=1}^{n} A_i\right) = \prod_{i=1}^{n} \mathrm{P}(A_i).$$

This is called the *multiplication rule* for independent events. Note that it is not a single condition involving only the product of all the probabilities of all single events (see below for a counterexample); it must hold true for all subset of events.

For more than two events, a mutually independent set of events is (by definition) pairwise independent; but the converse is not necessarily true (see below for a counterexample).

For Random Variables

Two Random Variables

Two random variables X and Y are **independent** if and only if (iff) the elements of the π-system generated by them are independent; that is to say, for every a and b, the events $\{X \leq a\}$ and $\{Y \leq b\}$ are independent events (as defined above). That is, X and Y with cumulative distribution functions $F_X(x)$ and $F_Y(y)$and $f_X(x)$ and $f_Y(y)$, , are independent iff the combined random variable (X, Y) has a joint cumulative distribution function

$$F_{X,Y}(x,y) = F_X(x)F_Y(y),$$

or equivalently, if the joint density exists,

$$f_{X,Y}(x,y) = f_X(x)f_Y(y).$$

More than two Random Variables

A set of random variables is **pairwise independent** if and only if every pair of random variables is independent.

A set of random variables is **mutually independent** if and only if for any finite subset $X_1,\ldots,X_n$ and any finite sequence of numbers $a_1,\ldots,a_n$, the events $\{X_1 \leq a_1\},\ldots,\{X_n \leq a_n\}$ are mutually independent events (as defined above).

The measure-theoretically inclined may prefer to substitute events $\{X \in A\}$ for events $\{X \leq a\}$ in the above definition, where A is any Borel set. That definition is exactly equivalent to the one above when the values of the random variables are real numbers. It has the advantage of working also for complex-valued random variables or for random variables taking values in any measurable space (which includes topological spaces endowed by appropriate σ-algebras).

Conditional Independence

Intuitively, two random variables X and Y are conditionally independent given Z if, once Z is known, the value of Y does not add any additional information about X. For instance, two measurements X and Y of the same underlying quantity Z are not independent, but they are **conditionally independent given Z** (unless the errors in the two measurements are somehow connected).

The formal definition of conditional independence is based on the idea of conditional distributions. If X, Y, and Z are discrete random variables, then we define X and Y to be *conditionally independent given Z* if

$$\mathrm{P}(X \leq x, Y \leq y \mid Z = z) = \mathrm{P}(X \leq x \mid Z = z) \cdot \mathrm{P}(Y \leq y \mid Z = z)$$

for all x, y and z such that $\mathrm{P}(Z = z) > 0$. On the other hand, if the random variables are continuous and have a joint probability density function p, then X and Y are conditionally independent given Z if

$$p_{XY|Z}(x,y|z) = p_{X|Z}(x|z) \cdot p_{Y|Z}(y|z)$$

for all real numbers x, y and z such that $p_Z(z) > 0$.

If X and Y are conditionally independent given Z, then

$$\mathrm{P}(X = x|Y = y, Z = z) = \mathrm{P}(X = x|Z = z)$$

for any x, y and z with $\mathrm{P}(Z = z) > 0$. That is, the conditional distribution for X given Y and Z is the same as that given Z alone. A similar equation holds for the conditional probability density functions in the continuous case.

Independence can be seen as a special kind of conditional independence, since probability can be seen as a kind of conditional probability given no events.

Independent σ-Algebras

The definitions above are both generalized by the following definition of independence for σ-algebras. Let (Ω, Σ, Pr) be a probability space and let **A** and **B** be two sub-σ-algebras of Σ. **A** and **B** are said to be **independent** if, whenever $A \in \mathbf{A}$ and $B \in \mathbf{B}$,

$$\mathrm{P}(A \cap B) = \mathrm{P}(A)\mathrm{P}(B).$$

Likewise, a finite family of σ-algebras $(\tau_i)_{i \in I}$ is said to be independent if and only if

$$\forall (A_i)_{i \in I} \in \prod_{i \in I} \tau_i \ : \ \mathrm{P}\left(\bigcap_{i \in I} A_i\right) = \prod_{i \in I} \mathrm{P}(A_i)$$

and an infinite family of σ-algebras is said to be independent if all its finite subfamilies are independent.

The new definition relates to the previous ones very directly:

- Two events are independent (in the old sense) if and only if the σ-algebras that they generate are independent (in the new sense). The σ-algebra generated by an event $E \in \Sigma$ is, by definition,

$$\sigma(\{E\}) = \{\emptyset, E, \Omega \setminus E, \Omega\}.$$

- Two random variables X and Y defined over Ω are independent (in the old sense) if and only if the σ-algebras that they generate are independent (in the new sense). The σ-algebra generated by a random variable X taking values in some measurable space S consists, by definition, of all subsets of Ω of the form $X^{-1}(U)$, where U is any measurable subset of S.

Using this definition, it is easy to show that if X and Y are random variables and Y is constant, then X and Y are independent, since the σ-algebra generated by a constant random variable is the trivial σ-algebra {∅, Ω}. Probability zero events cannot affect independence so independence also holds if Y is only Pr-almost surely constant.

PROPERTIES

Self-Independence

Note that an event is independent of itself if and only if

$$P(A) = P(A \cap A) = P(A) \cdot P(A) \Leftrightarrow P(A) = 0 \text{ or } 1.$$

Thus an event is independent of itself if and only if it almost surely occurs or its complement almost surely occurs.

Expectation and Covariance

If X and Y are independent, then the expectation operator E has the property

$$E[XY] = E[X]E[Y],$$

and the covariance cov(X, Y) is zero, since we have

$$\mathrm{cov}[X, Y] = E[XY] - E[X]E[Y].$$

(The converse of these, i.e. the proposition that if two random variables have a covariance of 0 they must be independent, is not true. See uncorrelated.)

Characteristic Function

Two random variables X and Y are independent if and only if the characteristic function of the random vector (X, Y) satisfies

$$\varphi_{(X,Y)}(t, s) = \varphi_X(t) \cdot \varphi_Y(s).$$

In particular the characteristic function of their sum is the product of their marginal characteristic functions:

$$\varphi_{X+Y}(t) = \varphi_X(t) \cdot \varphi_Y(t),$$

though the reverse implication is not true. Random variables that satisfy the latter condition are called subindependent.

EXAMPLES

Rolling Dice

The event of getting a 6 the first time a die is rolled and the event of getting a 6 the second time are *independent*. By contrast, the event of getting a 6 the first time a die is rolled and the event that the sum of the numbers seen on the first and second trials is 8 are *not* independent.

Drawing Cards

If two cards are drawn *with* replacement from a deck of cards, the event of drawing a red card on the first trial and that of drawing a red card on the second trial are *independent*. By contrast, if two cards are drawn *without* replacement from a deck of cards, the event of drawing a red card on the first trial and that of drawing a red card on the second trial are again *not* independent.

Pairwise and Mutual Independence

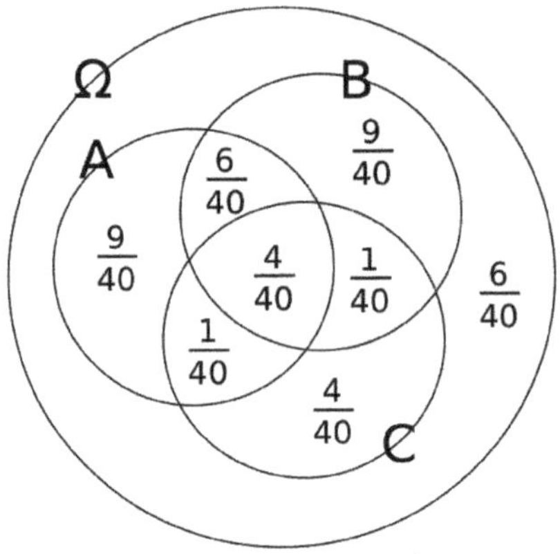

Pairwise independent, but not mutually independent, events.

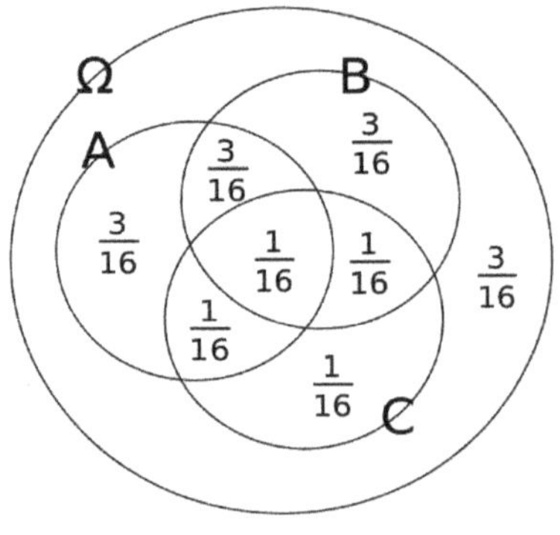

Mutually independent events.

Consider the two probability spaces shown. In both cases, $P(A) = P(B) = 1/2$ and $P(C) = 1/4$ The first space is pairwise independent but not mutually independent. The second space is both pairwise independent and mutually independent. To illustrate the difference, consider conditioning on two events. In the pairwise independent case, although any one event is independent of each of the other two individually, it is not independent of the intersection of the other two:

$$P(A|BC) = \frac{\frac{4}{40}}{\frac{4}{40} + \frac{1}{40}} = \tfrac{4}{5} \neq P(A)$$

$$P(B|AC) = \frac{\frac{4}{40}}{\frac{4}{40} + \frac{1}{40}} = \tfrac{4}{5} \neq P(B)$$

$$P(C|AB) = \frac{\frac{4}{40}}{\frac{4}{40} + \frac{6}{40}} = \tfrac{2}{5} \neq P(C)$$

In the mutually independent case however:

$$P(A|BC) = \frac{\frac{1}{16}}{\frac{1}{16} + \frac{1}{16}} = \tfrac{1}{2} = P(A)$$

$$P(B|AC) = \frac{\frac{1}{16}}{\frac{1}{16} + \frac{1}{16}} = \tfrac{1}{2} = P(B)$$

$$P(C|AB) = \frac{\frac{1}{16}}{\frac{1}{16} + \frac{3}{16}} = \tfrac{1}{4} = P(C)$$

Mutual Independence

It is possible to create a three-event example in which

$$P(A \cap B \cap C) = P(A)P(B)P(C),$$

and yet no two of the three events are pairwise independent (and hence the set of events are not mutually independent).This example shows that mutual independence involves requirements on the products of probabilities of all combinations of events, not just the single events as in this example.

CONDITIONAL INDEPENDENCE

In probability theory, two events R and B are **conditionally independent** given a third event Y precisely if the occurrence or non-occurrence of R *and* the occurrence or non-occurrence of B are independent events in their conditional probability distribution given Y. In other words, R and B are conditionally independent given Y if and only if, given knowledge that Y occurs, knowledge of whether R occurs provides no information on the likelihood of B occurring, and knowledge of whether B occurs provides no information on the likelihood of R occurring.

Formal Definition

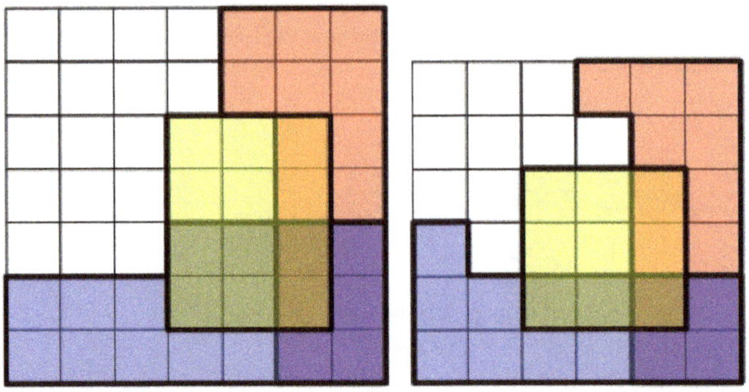

These are two examples illustrating **conditional independence**. Each cell represents a possible outcome. The events *R*, *B* and *Y* are represented by the areas shaded red, blue and yellow respectively. The overlap between the events R and B is shaded purple. The probabilities of these events are shaded areas with respect to the total area. In both examples *R* and *B* are conditionally independent given *Y* because: $\Pr(R \cap B \mid Y) = \Pr(R \mid Y)\Pr(B \mid Y)$ but not conditionally independent given not *Y* because: A, χ_A, given the sigma algebra Σ

In the standard notation of probability theory, *R* and *B* are conditionally independent given *Y* if and only if

$$\Pr(R \cap B \mid Y) = \Pr(R \mid Y)\Pr(B \mid Y),$$

or equivalently,

$$\Pr(R \mid B \cap Y) = \Pr(R \mid Y).$$

Two random variables *X* and *Y* are conditionally independent given a third random variable *Z* if and only if they are independent in their conditional probability distribution given *Z*. That is, *X* and *Y* are conditionally independent given *Z* if and only if, given any value of *Z*, the probability distribution of *X* is the same for all values of *Y* and the probability distribution of *Y* is the same for all values of *X*.

Two events *R* and *B* are conditionally independent given a σ-algebra Σ if

$$\Pr(R \cap B \mid \Sigma) = \Pr(R \mid \Sigma)\Pr(B \mid \Sigma) \ a.s.$$

where $\Pr(A \mid \Sigma)$ denotes the conditional expectation of the indicator function of the event A, χ_A, given the sigma algebra Σ. That is,

$$\Pr(A \mid \Sigma) := \mathrm{E}[\chi_A \mid \Sigma].$$

Two random variables *X* and *Y* are conditionally independent given a σ-algebra *Σ* if the above equation holds for all *R* in σ(*X*) and B in σ(*Y*).

Two random variables *X* and *Y* are conditionally independent given a random variable *W* if they are independent given σ(*W*): the σ-algebra generated by *W*. This is commonly written:

$X \perp\!\!\!\perp Y \mid W$ or
$X \perp Y \mid W$

This is read "X is independent of Y, given W"; the conditioning applies to the whole statement: "(X is independent of Y) given W".

If W assumes a countable set of values, this is equivalent to the conditional independence of X and Y for the events of the form $[W = w]$. Conditional independence of more than two events, or of more than two random variables, is defined analogously.

The following two examples show that $X \perp Y$ *neither implies nor is implied by* $X \perp Y \mid W$. First, suppose W is 0 with probability 0.5 and 1 otherwise. When $W = 0$ take X and Y to be independent, each having the value 0 with probability 0.99 and the value 1 otherwise. When $W = 1$, X and Y are again independent, but this time they take the value 1 with probability 0.99. Then $X \perp Y \mid W$. But X and Y are dependent, because $\Pr(X = 0) < \Pr(X = 0|Y = 0)$. This is because $\Pr(X = 0) = 0.5$, but if $Y = 0$ then it's very likely that $W = 0$ and thus that $X = 0$ as well, so $\Pr(X = 0|Y = 0) > 0.5$. For the second example, suppose $X \perp Y$, each taking the values 0 and 1 with probability 0.5. Let W be the product $X \times Y$. Then when $W = 0$, $\Pr(X = 0) = 2/3$, but $\Pr(X = 0|Y = 0) = 1/2$, so $X \perp Y \mid W$ is false. This is also an example of Explaining Away. See Kevin Murphy's tutorial where X and Y take the values "brainy" and "sporty".

An Example

The discussion on StackExchange provides a couple of useful examples.

- Let the two events be the probabilities of persons A and B getting home in time for dinner, and the third event is the fact that a snow storm hit the city. While both A and B have a lower probability of getting home in time for dinner, the lower probabilities will still be independent of each other. That is, the knowledge that A is late does not tell you whether B will be late. (They may be living in different neighborhoods, traveling different distances, and using different modes of transportation.) However, if you have information that they live in the same neighborhood, use the same transportation, and work at the same place, then the two events are NOT conditionally independent.
- Conditional independence depends on the nature of the third event. If you roll two dice, one may assume that the two dice behave independently of each other. Looking at the results of 1 die will not tell you about the result of the second die. (That is, the two dice are independent.) If, however, the 1st die`s result is a 3, and someone tells you about a third event - that the sum of the two results is even - then this extra unit of information restricts the options for the 2nd result to an odd number. In other words, two events can be independent, but NOT conditionally independent.
- Height and vocabulary are not independent; but they are conditionally independent if you add age.

Uses in Bayesian inference

Let p be the proportion of voters who will vote "yes" in an upcoming referendum. In taking an opinion poll, one chooses n voters randomly from the population. For $i = 1, ..., n$, let $X_i = 1$ or 0 corresponding, respectively, to whether or not the ith chosen voter will or will not vote "yes".

In a frequentist approach to statistical inference one would not attribute any probability distribution to p (unless the probabilities could be somehow interpreted as relative frequencies of occurrence of some event or as proportions of some population) and one would say that $X_1, ..., X_n$ are independent random variables.

By contrast, in a Bayesian approach to statistical inference, one would assign a probability distribution to p regardless of the non-existence of any such "frequency" interpretation, and one would construe the probabilities as degrees of belief that p is in any interval to which a probability is assigned. In that model, the random variables $X_1, ..., X_n$ are *not* independent, but they are **conditionally independent** given the value of p. In particular, if a large number of the Xs are observed to be equal to 1, that would imply a high conditional probability, given that observation, that p is near 1, and thus a high conditional probability, given that observation, that the *next* X to be observed will be equal to 1.

Rules of Conditional Independence

A set of rules governing statements of conditional independence have been derived from the basic definition.

Note: since these implications hold for any probability space, they will still hold if one considers a sub-universe by conditioning everything on another variable, say K. For example, $X \perp\!\!\!\perp Y \Rightarrow Y \perp\!\!\!\perp X$ would also mean that $X \perp\!\!\!\perp Y \mid K \Rightarrow Y \perp\!\!\!\perp X \mid K$. .

Note: below, the comma can be read as an "AND".

Symmetry

$$X \perp\!\!\!\perp Y \quad \Rightarrow \quad Y \perp\!\!\!\perp X$$

Decomposition

$$X \perp\!\!\!\perp A, B \quad \Rightarrow \quad \text{and} \begin{cases} X \perp\!\!\!\perp A \\ X \perp\!\!\!\perp B \end{cases}$$

Proof:

- $p_{X,A,B}(x, a, b) = p_X(x) p_{A,B}(a, b)$ (meaning of $X \perp\!\!\!\perp A, B$)
- $\int_B p_{X,A,B}(x, a, b) = \int_B p_X(x) p_{A,B}(a, b)$ (ignore variable B by integrating it out)
- $p_{X,A}(x, a) = p_X(x) p_A(a)$

A similar proof shows the independence of X and B.

Weak Union

$$X \perp\!\!\!\perp A,B \quad \Rightarrow \quad \text{and} \begin{cases} X \perp\!\!\!\perp A \mid B \\ X \perp\!\!\!\perp B \mid A \end{cases}$$

Proof:

- By definition, $\Pr(X) = \Pr(X \mid A, B)$.
- Due to the property of decomposition $X \perp\!\!\!\perp B$, $\Pr(X) = \Pr(X \mid B)$.
- Combining the above two equalities gives $\Pr(X \mid B) = \Pr(X \mid A, B)$, which establishes $X \perp\!\!\!\perp A \mid B$.

The second condition can be proved similarly.

Contraction

$$\left. \begin{matrix} X \perp\!\!\!\perp A \mid B \\ X \perp\!\!\!\perp B \end{matrix} \right\} \text{and} \quad \Rightarrow \quad X \perp\!\!\!\perp A, B$$

Proof:

This property can be proved by noticing $\Pr(X \mid A, B) = \Pr(X \mid B) = \Pr(X)$, each equality of which is asserted by $X \perp\!\!\!\perp A \mid B$ and $X \perp\!\!\!\perp B$, respectively.

Contraction-Weak-Union-Decomposition

Putting the above three together, we have:

$$\left. \begin{matrix} X \perp\!\!\!\perp A \mid B \\ X \perp\!\!\!\perp B \end{matrix} \right\} \text{and} \quad \iff \quad X \perp\!\!\!\perp A, B \quad \Rightarrow \quad \text{and} \begin{cases} X \perp\!\!\!\perp A \mid B \\ X \perp\!\!\!\perp B \\ X \perp\!\!\!\perp B \mid A \\ X \perp\!\!\!\perp A \end{cases}$$

Intersection

For strictly positive probability distributions, the following also holds:

$$\left. \begin{matrix} X \perp\!\!\!\perp A \mid C, B \\ X \perp\!\!\!\perp B \mid C, A \end{matrix} \right\} \text{and} \quad \Rightarrow \quad X \perp\!\!\!\perp B, A \mid C$$

The five rules above were termed "Graphoid Axioms" by Pearl and Paz, because they hold in graphs, if $X \perp\!\!\!\perp A \mid B$ is interpreted to mean: "All paths from X to A are intercepted by the set B".

GRAPHOID

A **graphoid** is a set of statements of the form, "X is irrelevant to Y given that we know Z" where X, Y and Z are sets of variables. The notion of "irrelevance" and "given that we know" may obtain different interpretations, including probabilistic, relational and correlational, depending on the application. These interpretations share common properties that can be captured by paths in graphs (hence the name "graphoid"). The theory of graphoids characterizes these properties in a finite set of axioms that are common to informational irrelevance and its graphical representations.

Judea Pearl and Azaria Paz coined the term "graphoids" after discovering that a set of axioms that govern conditional independence in probability theory is shared by undirected graphs. Variables are represented as nodes in a graph in such a way that variable sets X and Y are independent conditioned on Z in the distribution whenever node set Z separates X from Y in the graph. Axioms for conditional independence in probability were derived earlier by A. Philip Dawid and Wolfgang Spohn. The correspondence between dependence and graphs was later extended to directed acyclic graphs (DAGs) and to other models of dependency.

A dependency model M is a subset of triplets (X,Z,Y) for which the predicate $I(X,Z,Y)$: X is independent of Y given Z, is true. A graphoid is defined as a dependency model that is closed under the following five axioms:

1. Symmetry: $I(X,Z,Y) \Leftrightarrow I(Y,Z,X)$
2. Decomposition: $I(X,Z,Y \cup W) \Rightarrow I(X,Z,Y) \,\&\, I(X,Z,W)$
3. Weak Union: $I(X,Z,Y \cup W) \Rightarrow I(X,Z \cup W,Y)$
4. Contraction: $I(X,Z,Y) \,\&\, I(X,Z \cup Y,W) \Rightarrow I(X,Z,Y \cup W)$
5. Intersection: $I(X,Z \cup W,Y) \,\&\, I(X,Z \cup Y,W) \Rightarrow I(X,Z,Y \cup W)$

A semi-graphoid is a dependency model closed under (i)–(iv). These five axioms together are known as the graphoid axioms. Intuitively, the weak union and contraction properties mean that irrelevant information should not alter the relevance status of other propositions in the system; what was relevant remains relevant and what was irrelevant remains irrelevant.

TYPES OF GRAPHOIDS

Probabilistic Graphoids

Conditional independence, defined as

$$I(X,Z,Y) \text{ iff } P(X \mid Y,Z) = P(X \mid Z)$$

is a semi-graphoid which becomes a full graphoid when P is strictly positive.

Correlational Graphoids

A dependency model is a correlational graphoid if in some probability function we have,

$$I_c(X,Y,Z) \Leftrightarrow \rho_{xy.z} = 0 \text{ for every } x \in X \text{ and } y \in Y$$

where $\rho_{xy.z}$ is the partial correlation between x and y given set Z.

In other words, the linear estimation error of the variables in X using measurements on Z would not be reduced by adding measurements of the variables in Y, thus making Y irrelevant to the estimation of X. Correlational and probabilistic dependency models coincide for normal distributions.

Relational Graphoids

A dependency model is a relational graphoid if it satisfies

$$P(X,Z) > 0 \,\&\, P(Y,Z) > 0 \implies P(X,Y,Z) > 0.$$

In words, the range of values permitted for X is not restricted by the choice of Y, once Z is fixed. Independence statements belonging to this model are similar to embedded multi-valued dependencies (EMVD s) in databases.

Graph-Induced Graphoids

If there exists an undirected graph G such that,

$$I(X,Z,Y) \Leftrightarrow \langle X,Z,Y\rangle_G,$$

then the graphoid is called as graph-induced. In other words, there exists an undirected graph G such that every independence statement in M is reflected as a vertex separation in G and vice versa. A necessary and sufficient condition for a dependency model to be a graph induced graphoid is that it satisfies the following axioms: symmetry, decomposition, intersection, strong union and transitivity.

Strong union states that,

$$I(X,Z,Y) \implies I(X,Z \cup W,Y)$$

Transitivity states that

$$I(X,Z,Y) \implies I(X,Z,\gamma) \text{ or } I(\gamma,Z,Y) \;\forall\; \gamma \notin X \cup Y \cup Z.$$

Graphoid axioms constitute a complete characterization of undirected graphs.

DAG-Induced Graphoids

A graphoid is termed DAG-induced if there exists a directed acyclic graph D such that $I(X,Z,Y) \Leftrightarrow \langle X,Z,Y\rangle_D$ where $\langle X,Z,Y\rangle_D$ stands for d-separation in D. d-separation (d-connotes "directional") extends the notion of vertex separation from undirected graphs to directed acyclic graphs. It permits the reading of conditional independencies from the structure of

Bayesian networks. However, conditional independencies in a DAG cannot be completely characterized by a finite set of axioms.

Inclusion and Construction

Graph-induced and DAG-induced graphoids are both contained in probabilistic graphoids. This means that for every graph *G* there exists a probability distribution *P* such that every conditional independence in *P* is represented in *G*, and vice versa. The same is true for DAGs. However, there are probabilistic distributions that are not semi-graphoids and, moreover, there is no finite axiomatization for probabilistic dependencies.

Thomas Verma showed that every semi-graphoid has a recursive way of constructing a DAG in which every *d*-separation is valid. The construction is similar to that used in Bayes networks and goes as follows:

- Arrange the variables in some arbitrary order 1, 2,...,i,...,*N* and, starting with $i = 1$,
- choose for each node *i* a set of nodes PA_i such that *i* is independent on all its predecessors, 1, 2,...,$i - 1$, conditioned on PA_i.
- Draw arrows from PA_i to *i* and continue.

The DAG created by this construction will represent all the conditional independencies that follow from those used in the construction. Furthermore, every *d*-separation shown in the DAG will be a valid conditional independence in the graphoid used in the construction.

REFERENCES

1. Pearl, Judea; Paz, Azaria (1985). "Graphoids: A Graph-Based Logic for Reasoning About Relevance Relations".
2. Dawid, A. Philip (1979). "Conditional independence in statistical theory". Journal of the Royal Statistical Society, Series B (Methodological): 1–31.
3. Spohn, Wolfgang (1980). "Stochastic independence, causal independence, and shieldability". Journal of Philosophical Logic. 9: 73–99. doi:10.1007/bf00258078.
4. Pearl, Judea (1986). "Fusion, propagation and structuring in belief networks". Artificial Intelligence. 29 *(3): 241–288.* doi:10.1016/0004-3702(86)90072-x.
5. Verma, Thomas; Pearl, Judea (1988). "Causal networks: Semantics and expressiveness". Proceedings of the 4th Workshop on Uncertainty in Artificial Intelligence: 352–359.
6. Lauritzen, S.L. (1996). Graphical Models. Oxford: Clarendon Press.
7. Geiger, Dan (1990). "Graphoids: A Qualitative Framework for Probabilistic Inference" (PhD Dissertation, Technical Report R-142, Computer Science Department, University of California, Los Angeles). line feed character in |format= at position 59
8. Pearl, Judea (1988). Probabilistic reasoning in intelligent systems: networks of plausible inference. Morgan Kaufmann.

9. A. Paz, J. Pearl, and S. Ur, "A New Characterization of Graphs Based on Interception Relations" Journal of Graph Theory, Vol. 22, No. 2, 125-136, 1996.

10. Geiger, D. (1987). "The non-axiomatizability of dependencies in directed acyclic graphs" (PDF). UCLA Computer Science Tech Report R-83.

11. Geiger, D.; Pearl, J. (1993). "Logical and algorithmic properties of conditional independence and graphical models". The Annals of Statistics. 21 *(4): 2001–2021. doi:10.1214/*aos/1176349407.

12. Studeny, M. (1992). Kubik, S.; Visek, J.A., eds. "Conditional independence relations have no finite complete characterization". Information Theory, Statistical Decision Functions and Random Processes. Transactions of the 11th Prague Conference. Dordrecht: Kluwer. B: 377–396.

13. Verma, T.; Pearl, J. (1990). Shachter, R.; Levitt, T.S.; Kanal, L.N., eds. "Causal Networks: Semantics and Expressiveness". Uncertainty in AI 4. Elsevier Science Publishers: 69–76.

Law of Total Probability

Chapter Contents

In probability theory, the **law** (or **formula**) **of total probability** is a fundamental rule relating marginal probabilities to conditional probabilities. It expresses the total probability of an outcome which can be realized via several distinct events.

STATEMENT

The law of total probability is the proposition that if $\{B_n : n = 1, 2, 3, \ldots\}$ is a finite or countably infinite partition of a sample space (in other words, a set of pairwise disjoint events whose union is the entire sample space) and each event B_n is measurable, then for any event A of the same probability space:

$$\Pr(A) = \sum_n \Pr(A \cap B_n)$$

or, alternatively,

$$\Pr(A) = \sum_n \Pr(A \mid B_n) \Pr(B_n),$$

where, for any n for which $\Pr(B_n) = 0$ these terms are simply omitted from the summation, because $\Pr(A \mid B_n)$ is finite.

The summation can be interpreted as a weighted average, and consequently the marginal probability, $\Pr(A)$, is sometimes called "average probability"; "overall probability" is sometimes used in less formal writings.

The law of total probability can also be stated for conditional probabilities. Taking the B_n as above, and assuming C is an event independent with any of the B_n:

$$\Pr(A \mid C) = \sum_n \Pr(A \mid C \cap B_n) \Pr(B_n \mid C) = \sum_n \Pr(A \mid C \cap B_n) \Pr(B_n)$$

INFORMAL FORMULATION

The above mathematical statement might be interpreted as follows: *given an outcome A, with known conditional probabilities given any of the B_n events, each with a known probability itself, what is the total probability that A will happen?* The answer to this question is given by $\Pr(A)$.

EXAMPLE

Suppose that two factories supply light bulbs to the market. Factory *X*'s bulbs work for over 5000 hours in 99% of cases, whereas factory *Y*'s bulbs work for over 5000 hours in 95% of cases. It is known that factory *X* supplies 60% of the total bulbs available. What is the chance that a purchased bulb will work for longer than 5000 hours?

Applying the law of total probability, we have:

$$\Pr(A) = \Pr(A|B_X) \cdot \Pr(B_X) + \Pr(A|B_Y) \cdot \Pr(B_Y) = \frac{99}{100} \cdot \frac{6}{10} + \frac{95}{100} \cdot \frac{4}{10} = \frac{594 + 380}{1000} = \frac{974}{1000}$$

where

- $\Pr(B_X) = \frac{6}{10}$ is the probability that the purchased bulb was manufactured by factory *X*;
- $\Pr(B_Y) = \frac{4}{10}$ is the probability that the purchased bulb was manufactured by factory *Y*;
- $\Pr(A|B_X) = \frac{99}{100}$ is the probability that a bulb manufactured by *X* will work for over 5000 hours;
- $\Pr(A|B_Y) = \frac{95}{100}$ is the probability that a bulb manufactured by *Y* will work for over 5000 hours.

Thus each purchased light bulb has a 97.4% chance to work for more than 5000 hours.

APPLICATIONS

One common application of the law is where the events coincide with a discrete random variable *X* taking each value in its range, i.e B_n is the event $X = x_n$. It follows that the probability of an event *A* is equal to the expected value of the conditional probabilities of *A* given

$X = x_n$. That is,

$$\Pr(A) = \sum_n \Pr(A \mid X = x_n) \Pr(X = x_n) = \mathrm{E}[\Pr(A \mid X)],$$

where Pr(*A* | *X*) is the conditional probability of *A* given the value of the random variable *X*. This conditional probability is a random variable in its own right, whose value depends on that of *X*. The conditional probability Pr(*A* | *X* = x) is simply a conditional probability given an event, [*X* = *x*]. It is a function of *x*, say *g*(*x*) = Pr(*A* | *X* = *x*). Then the conditional probability Pr(*A* | *X*) is *g*(*X*), hence itself a random variable. This version of the law of total probability says that the expected value of this random variable is the same as Pr(*A*).

This result can be generalized to continuous random variables (via continuous conditional density), and the expression becomes

$$\Pr(A) = \mathrm{E}[\Pr(A \mid \mathcal{F}_X)],$$

where $\mathcal{F}_X$ denotes the sigma-algebra generated by the random variable *X*.

OTHER NAMES

The term ***law of total probability*** is sometimes taken to mean the **law of alternatives**, which is a special case of the law of total probability applying to discrete random variables. One author even uses the terminology "continuous law of alternatives" in the continuous case. This result is given by Grimmett and Welsh as the **partition theorem**, a name that they also give to the related law of total expectation.

LAW OF TOTAL VARIANCE

From Wikipedia, the free encyclopedia

In probability theory, the **law of total variance** or **variance decomposition formula**, also known as **Eve's law**, states that if *X* and *Y* are random variables on the same probability space, and the variance of *Y* is finite, then

$$\mathrm{Var}(Y) = \mathrm{E}[\mathrm{Var}(Y|X)] + \mathrm{Var}(\mathrm{E}[Y|X]).$$

Some writers on probability call this the "conditional variance formula". In language perhaps better known to statisticians than to probabilists, the two terms are the "unexplained" and the "explained" components of the variance respectively (cf. fraction of variance unexplained, explained variation). In actuarial science, specifically credibility theory, the first component is called the expected value of the process variance (**EVPV**) and the second is called the variance of the hypothetical means (**VHM**).

There is a general variance decomposition formula for $c \geq 2$ components (see below). For example, with two conditioning random variables:

$$\mathrm{Var}[Y] = \mathrm{E}[\mathrm{Var}(Y|X_1, X_2)] + \mathrm{E}[\mathrm{Var}(\mathrm{E}[Y|X_1, X_2]|X_1)] + \mathrm{Var}(\mathrm{E}[Y|X_1]),$$

which follows from the law of total conditional variance:

$$\mathrm{Var}(Y|X_1) = \mathrm{E}[\mathrm{Var}(Y|X_1, X_2)|X_1] + \mathrm{Var}(\mathrm{E}[Y|X_1, X_2]|X_1).$$

Note that the conditional expected value E(Y | X) is a random variable in its own right, whose value depends on the value of X. Notice that the conditional expected value of Y given the *event* $X = x$ is a function of x (this is where adherence to the conventional and rigidly case-sensitive notation of probability theory becomes important!). If we write E(Y | $X = x$) = $g(x)$ then the random variable E(Y | X) is just $g(X)$. Similar comments apply to the conditional variance.

One special case, (similar to the Law of total expectation) states that if $A_1, \cdots, A_n$ is a partition of the whole outcome space, i.e. these events are mutually exclusive and exhaustive, then

$$\mathrm{Var}(X) = \sum_{i=1}^{n} \mathrm{Var}(X|A_i)\Pr(A_i) + \sum_{i=1}^{n} \mathrm{E}[X|A_i]^2(1 - \Pr(A_i))\Pr(A_i) - 2\sum_{i=2}^{n}\sum_{j=1}^{i-1} \mathrm{E}[X|A_i]\Pr(A_i)\mathrm{E}[X|A_j]\Pr(A_j).$$

In this formula, the first component is the expectation of the conditional variance; the other two rows are the variance of the conditional expectation.

Proof

The law of total variance can be proved using the law of total expectation. First,

$$\mathrm{Var}[Y] = \mathrm{E}[Y^2] - \mathrm{E}[Y]^2$$

from the definition of variance. Then we apply the law of total expectation to each term by conditioning on the random variable X:

$$= \mathrm{E}\left[\mathrm{E}[Y^2 \mid X]\right] - [\mathrm{E}[\mathrm{E}[Y \mid X]]]^2$$

Now we rewrite the conditional second moment of Y in terms of its variance and first moment:

$$= \mathrm{E}\left[\mathrm{Var}[Y \mid X] + [\mathrm{E}[Y \mid X]]^2\right] - [\mathrm{E}[\mathrm{E}[Y \mid X]]]^2$$

Since the expectation of a sum is the sum of expectations, the terms can now be regrouped:

$$= \mathrm{E}[\mathrm{Var}[Y \mid X]] + \left(\mathrm{E}[[\mathrm{E}[Y \mid X]]^2] - [\mathrm{E}[\mathrm{E}[Y \mid X]]]^2\right)$$

Finally, we recognize the terms in parentheses as the variance of the conditional expectation E[Y|X]:

$$= \mathrm{E}[\mathrm{Var}[Y \mid X]] + \mathrm{Var}[\mathrm{E}[Y \mid X]]$$

General variance decomposition applicable to dynamic systems

The following formula shows how to apply the general, measure theoretic variance decomposition formula to stochastic dynamic systems. Let $Y(t)$ be the value of a system variable at time t. Suppose we have the internal histories (natural filtrations) $H_{1t}, H_{2t}, \ldots, H_{c-1,t}$, , each one corresponding to the history (trajectory) of a different collection of system variables. The collections need not be disjoint. The variance of $Y(t)$ can be decomposed, for all times t, into $c \geq 2$ components as follows:

$$\mathrm{Var}[Y(t)] = \mathrm{E}(\mathrm{Var}[Y(t) \mid H_{1t}, H_{2t}, \ldots, H_{c-1,t}]) + \sum_{j=2}^{c-1} \mathrm{E}(\mathrm{Var}[\mathrm{E}[Y(t) \mid H_{1t}, H_{2t}, \ldots, H_{jt}] \mid H_{1t}, H_{2t}, \ldots, H_{j-1,t}]) + \mathrm{Var}(\mathrm{E}[Y(t) \mid H_{1t}]).$$

The decomposition is not unique. It depends on the order of the conditioning in the sequential decomposition.

The Square of the Correlation and Explained (or Informational) Variation

In cases where (Y, X) are such that the conditional expected value is linear; i.e., in cases where

$$\mathrm{E}(Y \mid X) = aX + b,$$

it follows from the bilinearity of covariance that

$$a = \frac{\mathrm{Cov}(Y, X)}{\mathrm{Var}(X)}$$

and

$$b = \mathrm{E}(Y) - \frac{\mathrm{Cov}(Y, X)}{\mathrm{Var}(X)}\mathrm{E}(X)$$

and the explained component of the variance divided by the total variance is just the square of the correlation between Y and X; i.e., in such cases,

$$\frac{\mathrm{Var}(\mathrm{E}(Y \mid X))}{\mathrm{Var}(Y)} = \mathrm{Corr}(X, Y)^2.$$

One example of this situation is when (X, Y) have a bivariate normal (Gaussian) distribution.

More generally, when the conditional expectation $\mathrm{E}(Y \mid X)$ is a non-linear function of X

$$\iota_{Y|X} = \frac{\mathrm{Var}(\mathrm{E}(Y \mid X))}{\mathrm{Var}(Y)} = \mathrm{Corr}(\mathrm{E}(Y \mid X), Y)^2,$$

which can be estimated as the R squared from a non-linear regression of Y on X, using data drawn from the joint distribution of (X,Y). When $\mathrm{E}(Y \mid X)$ has a Gaussian distribution (and is an invertible function of X), or Y itself has a (marginal) Gaussian distribution, this explained component of variation sets a lower bound on the mutual information: $\mathrm{I}(Y; X) \geq \ln([1 - \iota_{Y|X}]^{-1/2})$.

Higher Moments

A similar law for the third central moment μ_3 says

$$\mu_3(Y) = \mathrm{E}(\mu_3(Y \mid X)) + \mu_3(\mathrm{E}(Y \mid X)) + 3\,\mathrm{cov}(\mathrm{E}(Y \mid X), \mathrm{var}(Y \mid X)).$$

For higher cumulants, a simple and elegant generalization exists. See law of total cumulance.

LAW OF TOTAL EXPECTATION

The proposition in probability theory known as the **law of total expectation**, the **law of iterated expectations**, the **tower rule**, the **smoothing theorem**, and **Adam's Law** among other names, states that if X is an integrable random variable (i.e., a random variable satisfying $\mathrm{E}(\mid X \mid < \infty)$ and Y is any random variable, not necessarily integrable, on the same probability space, then

$$\mathrm{E}(X) = \mathrm{E}(\mathrm{E}(X \mid Y)),$$

i.e., the expected value of the conditional expected value of X given Y is the same as the expected value of X.

The conditional expected value $\mathrm{E}(X \mid Y)$ is a random variable in its own right, whose value depends on the value of Y. Notice that the conditional expected value of X given the *event* $Y = y$ is a function of y. If we write $\mathrm{E}(X \mid Y = y) = g(y)$ then the random variable $\mathrm{E}(X \mid Y)$ is just $g(Y)$.

One special case states that if $A_1, A_2, \ldots, A_n$ is a partition of the whole outcome space, i.e. these events are mutually exclusive and exhaustive, then

$$\mathrm{E}(X) = \sum_{i=1}^{n} \mathrm{E}(X \mid A_i)\,\mathrm{P}(A_i).$$

Example

Suppose that two factories supply light bulbs to the market. Factory X's bulbs work for an average of 5000 hours, whereas factory Y's bulbs work for an average of 4000 hours. It is known that factory X supplies 60% of the total bulbs available. What is the expected length of time that a purchased bulb will work for?

Applying the law of total expectation, we have:

$$\mathrm{E}(L) = \mathrm{E}(L \mid X)\,\Pr(X) + \mathrm{E}(L \mid Y)\,\Pr(Y) = 5000(.6) + 4000(.4) = 4600$$

where

- $\mathbf{E}(L)$ is the expected life of the bulb;
- $\mathbf{Pr}(X) = \frac{6}{10}$ is the probability that the purchased bulb was manufactured by factory X;
- $\mathbf{Pr}(Y) = \frac{4}{10}$ is the probability that the purchased bulb was manufactured by factory Y;
- $\mathbf{E}(L \mid X) = 5000$ is the expected lifetime of a bulb manufactured by X;
- $\mathbf{E}(L \mid Y) = 4000$ is the expected lifetime of a bulb manufactured by Y.

Thus each purchased light bulb has an expected lifetime of 4600 hours.

Proof in the Discrete Case

$$\begin{aligned}
\mathrm{E}_Y\left(\mathrm{E}_{X|Y}(X \mid Y)\right) &= \mathrm{E}_Y\left[\sum_x x \cdot \mathrm{P}(X=x \mid Y)\right] \\
&= \sum_y \left[\sum_x x \cdot \mathrm{P}(X=x \mid Y=y)\right] \cdot \mathrm{P}(Y=y) \\
&= \sum_y \sum_x x \cdot \mathrm{P}(X=x \mid Y=y) \cdot \mathrm{P}(Y=y) \\
&= \sum_x x \sum_y \mathrm{P}(X=x \mid Y=y) \cdot \mathrm{P}(Y=y) \\
&= \sum_x x \sum_y \mathrm{P}(X=x, Y=y) \\
&= \sum_x x \cdot \mathrm{P}(X=x) \\
&= \mathrm{E}(X).
\end{aligned}$$

Proof in the General Case

The general statement of the result makes reference to a probability space $(\Omega,\mathcal{F},P)$ on which two sub {\displaystyle \sigma }-algebras $\mathcal{G}_1 \subseteq \mathcal{G}_2 \subseteq \mathcal{F}$ are defined. For a random variable X on such a space, the smoothing law states that $\mathbf{E}[\mathbf{E}[X \mid \mathcal{G}_2] \mid \mathcal{G}_1] = \mathbf{E}[X \mid \mathcal{G}_1]$.

Since a conditional expectation is a Radon–Nikodym derivative, verifying the following two properties establishes the smoothing law:

- $\mathbf{E}[\mathbf{E}[X \mid \mathcal{G}_2] \mid \mathcal{G}_1]$ **is** $\mathcal{G}_1$-measurable
- $\int_{G_1} \mathbf{E}[\mathbf{E}[X \mid \mathcal{G}_2] \mid \mathcal{G}_1] dP = \int_{G_1} X dP$ **holds for all** $G_1 \in \mathcal{G}_1$

The first of these properties holds by the definition of the conditional expectation, and the second holds since $G_1 \in \mathcal{G}_1 \subseteq \mathcal{G}_2$ implies

$$\int_{G_1} \mathrm{E}[\mathrm{E}[X \mid \mathcal{G}_2] \mid \mathcal{G}_1] dP = \int_{G_1} \mathrm{E}[X \mid \mathcal{G}_2] dP = \int_{G_1} X dP.$$

In the special case that $\mathcal{G}_1 = \{\emptyset, \Omega\}$ and $\mathcal{G}_2 = \sigma(Y)$, the smoothing law reduces to the statement

$$\mathrm{E}[\mathrm{E}[X \mid Y]] = \mathrm{E}[X].$$

NOTATION WITHOUT INDICES

When using the expectation operator E adding indices to the operator may lead to cumbersome notations and these indices are often omitted. In the case of iterated expectations $\mathrm{E}(\mathrm{E}(X \mid Y))$ stands for $\mathrm{E}_Y(\mathrm{E}_{X|Y}(X \mid Y))$. The innermost expectation is the conditional expectation of X given Y, and the outermost expectation is taken with respect to the conditioning variable Y.

Iterated Expectations with Nested Conditioning Sets

The following formulation of the **law of iterated expectations** plays an important role in many economic and finance models:

$$\mathrm{E}(X \mid I_1) = \mathrm{E}(\mathrm{E}(X \mid I_2) \mid I_1),$$

where the value of I_2 is determined by that of I_1. To build intuition, imagine an investor who forecasts a random stock price X based on the limited information set I_1. The law of iterated expectations says that the investor can never gain a more precise forecast of X by conditioning on more specific information (I_2), *if* the more specific forecast must itself be forecast with the original information (I_1).

This formulation is often applied in a time series context, where E_t denotes expectations conditional on only the information observed up to and including time period t. In typical models the information set $t + 1$ contains all information available through time t, plus additional information revealed at time $t + 1$. One can then write:

$$\mathrm{E}_t(X) = \mathrm{E}_t(\mathrm{E}_{t+1}(X)).$$

LAW OF TOTAL COVARIANCE

In probability theory, the **law of total covariance**, **covariance decomposition formula**, or **ECCE** states that if X, Y, and Z are random variables on the same probability space, and the covariance of X and Y is finite, then

$$\operatorname{cov}(X, Y) = \mathrm{E}(\operatorname{cov}(X, Y \mid Z)) + \operatorname{cov}(\mathrm{E}(X \mid Z), \mathrm{E}(Y \mid Z)).$$

The nomenclature in this article's title parallels the phrase *law of total variance*. Some writers on probability call this the "conditional covariance formula" or use other names.

(The conditional expected values E($X \mid Z$) and E($Y \mid Z$) are random variables in their own right, whose values depends on the value of Z. Notice that the conditional expected value of X given the *event* $Z = z$ is a function of z (this is where adherence to the conventional rigidly case-sensitive notation of probability theory becomes important!). If we write E($X \mid Z = z$) = $g(z)$ then the random variable E($X \mid Z$) is just $g(Z)$. Similar comments apply to the conditional covariance.)

Proof

The law of total covariance can be proved using the law of total expectation: First,
$$\operatorname{cov}[X,Y] = \mathrm{E}[XY] - \mathrm{E}[X]\,\mathrm{E}[Y]$$

from the definition of covariance. Then we apply the law of total expectation by conditioning on the random variable Z:
$$= \mathrm{E}[\mathrm{E}[XY \mid Z]] - \mathrm{E}[\mathrm{E}[X \mid Z]]\,\mathrm{E}[\mathrm{E}[Y \mid Z]]$$

Now we rewrite the term inside the first expectation using the definition of covariance:
$$= \mathrm{E}[\operatorname{cov}[X,Y \mid Z] + \mathrm{E}[X \mid Z]\,\mathrm{E}[Y \mid Z]] - \mathrm{E}[\mathrm{E}[X \mid Z]]\,\mathrm{E}[\mathrm{E}[Y \mid Z]]$$

Since expectation of a sum is the sum of expectations, we can regroup the terms:
$$= \mathrm{E}[\operatorname{cov}[X,Y \mid Z]] + \mathrm{E}[\mathrm{E}[X \mid Z]\,\mathrm{E}[Y \mid Z]] - \mathrm{E}[\mathrm{E}[X \mid Z]]\,\mathrm{E}[\mathrm{E}[Y \mid Z]]$$

Finally, we recognize the final two terms as the covariance of the conditional expectations E[$X|Z$] and E[$Y|Z$]:

REFERNCES

1. "Probability". *Webster's Revised Unabridged Dictionary*. G & C Merriam, 1913
2. Strictly speaking, a value of 0 indicates impossibility only when a sample space is finite. In an infinite sample space, a probability of 0 may indicate infinitesimal likelihood, but not impossibility. For example, the probability of any specific integer, say 5, being chosen completely at random from the entire set of integers is 0. However, it is not impossible for 5 to be chosen, just extremely unlikely.
3. "Kendall's Advanced Theory of Statistics, Volume 1: Distribution Theory", Alan Stuart and Keith Ord, 6th Ed, (2009), ISBN 9780534243128
4. William Feller, An Introduction to Probability Theory and Its Applications , (Vol 1), 3rd Ed, (1968), Wiley, ISBN 0-471-25708-7
5. Probability Theory The Britannica website
6. Hacking, Ian (1965). The Logic of Statistical Inference. Cambridge University Press. ISBN 0-521-05165-7.
7. Finetti, Bruno de (1970). Logical foundations and measurement of subjective probability . Acta Psychologica. 34: 129–145. doi:10.1016/0001-6918(70)90012-0.

8. Hájek, Alan. Interpretations of Probability . The Stanford Encyclopedia of Philosophy (Winter 2012 Edition), Edward N. Zalta (ed.). Retrieved 22 April 2013.
9. Hogg, Robert V.; Craig, Allen; McKean, Joseph W. (2004). Introduction to Mathematical Statistics (6th ed.). Upper Saddle River: Pearson. ISBN 0-13-008507-3.
10. Hacking, I. (2006) The Emergence of Probability: A Philosophical Study of Early Ideas about Probability, Induction and Statistical Inference, Cambridge University Press, ISBN 978-0-521-68557-3
11. Freund, John. (1973) Introduction to Probability. Dickenson ISBN 978-0822100782 (p. 1)
12. Jeffrey, R.C., Probability and the Art of Judgment, Cambridge University Press. (1992). pp. 54-55 . ISBN 0-521-39459-7
13. Franklin, J. (2001) The Science of Conjecture: Evidence and Probability Before Pascal, Johns Hopkins University Press. (pp. 22, 113, 127)
14. Some laws and problems in classical probability and how Cardano anticipated them Gorrochum, P. Chance magazine 2012
15. Abrams, William, A Brief History of Probability, Second Moment, retrieved 2008-05-23
16. Ivancevic, Vladimir G.; Ivancevic, Tijana T. (2008). Quantum leap : from Dirac and Feynman, across the universe, to human body and mind. Singapore ; Hackensack, NJ: World Scientific. p. 16. ISBN 978-981-281-927-7.
17. Franklin, James (2001). The Science of Conjecture: Evidence and Probability Before Pascal. Johns Hopkins University Press. ISBN 0801865697.
18. Wilson EB (1923) “First and second laws of error”. Journal of the American Statistical Association, 18, 143
19. Seneta, Eugene William. Adrien-Marie Legendre (version 9) . StatProb: The Encyclopedia Sponsored by Statistics and Probability Societies. Retrieved 27 January 2016.
20. http://www.statslab.cam.ac.uk/~rrw1/markov/M.pdf
21. Vitanyi, Paul M.B. (1988). Andrei Nikolaevich Kolmogorov . CWI Quarterly (1): 3–18. Retrieved 27 January 2016.
22. Singh, Laurie (2010) “Whither Efficient Markets? Efficient Market Theory and Behavioral Finance”. The Finance Professionals’ Post, 2010.
23. Gao, J.Z.; Fong, D.; Liu, X. (April 2011). Mathematical analyses of casino rebate systems for VIP gambling . International Gambling Studies. 11 (1): 93–106. doi:10.1080/14459795.2011.552575.
24. Gorman, Michael (2011) Management Insights . Management Science
25. Ross, Sheldon. A First course in Probability, 8th Edition. Page 26-27.
26. Olofsson (2005) Page 8.

27. Olofsson (2005), page 9

28. Olofsson (2005) page 35.

29. Olofsson (2005) page 29.

30. Burgi, Mark (2010) "Interpretations of Negative Probabilities", p. 1. arXiv:1008.1287v1

31. Jedenfalls bin ich überzeugt, daß der Alte nicht würfelt. Letter to Max Born, 4 December 1926, in: Einstein/Born Briefwechsel 1916-1955.

32. Moore, W.J. (1992). Schrödinger: Life and Thought. Cambridge University Press. p. 479. ISBN 0-521-43767-9.

CHAPTER 7

Law of Large Numbers

Chapter Contents

- **Introduction**
- **Examples**
- **History**
- **Forms**
- **Central Limit Theorem**

INTRODUCTION

In probability theory, the law of large numbers (LLN) is a theorem that describes the result of performing the same experiment a large number of times. According to the law, the average of the results obtained from a large number of trials should be close to the expected value, and will tend to become closer as more trials are performed.

The LLN is important because it "guarantees" stable long-term results for the averages of some random events. For example, while a casino may lose money in a single spin of the roulette wheel, its earnings will tend towards a predictable percentage over a large number of spins. Any winning streak by a player will eventually be overcome by the parameters of the game. It is important to remember that the LLN only applies (as the name indicates) when a large number of observations are considered. There is no principle that a small number of observations will coincide with the expected value or that a streak of one value will immediately be "balanced" by the others (see the gambler`s fallacy).

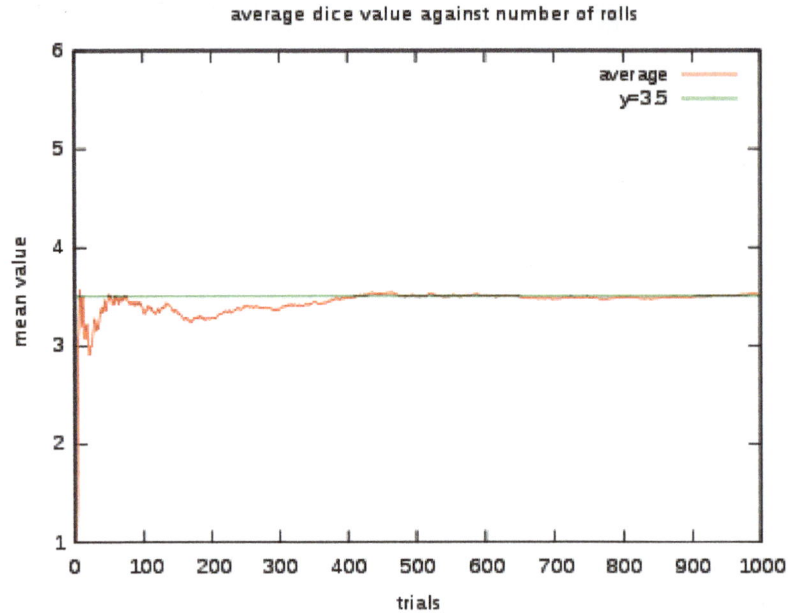

An illustration of the law of large numbers using a particular run of rolls of a single dice. As the number of rolls in this run increases, the average of the values of all the results approaches 3.5. While different runs would show a different shape over a small number of throws (at the left), over a large number of rolls (to the right) they would be extremely similar.

EXAMPLES

For example, a single roll of a fair, six-sided dice produces one of the numbers 1, 2, 3, 4, 5, or 6, each with equal probability. Therefore, the expected value of a single dice roll is

$$\frac{1+2+3+4+5+6}{6} = 3.5$$

According to the law of large numbers, if a large number of six-sided dice are rolled, the average of their values (sometimes called the sample mean) is likely to be close to 3.5, with the precision increasing as more dice are rolled.

It follows from the law of large numbers that the empirical probability of success in a series of Bernoulli trials will converge to the theoretical probability. For a Bernoulli random variable, the expected value is the theoretical probability of success, and the average of n such variables (assuming they are independent and identically distributed (i.i.d.)) is precisely the relative frequency.

For example, a fair coin toss is a Bernoulli trial. When a fair coin is flipped once, the theoretical probability that the outcome will be heads is equal to 1/2. Therefore, according to the law of large numbers, the proportion of heads in a large" number of coin flips should be" roughly 1/2. In particular, the proportion of heads after n flips will almost surely converge to 1/2 as n approaches infinity.

Though the proportion of heads (and tails) approaches 1/2, almost surely the absolute difference in the number of heads and tails will become large as the number of flips becomes large. That is, the probability that the absolute difference is a small number, approaches zero as the number of flips becomes large. Also, almost surely the ratio of the absolute difference to the number of flips will approach zero. Intuitively, expected absolute difference grows, but at a slower rate than the number of flips, as the number of flips grows.

HISTORY

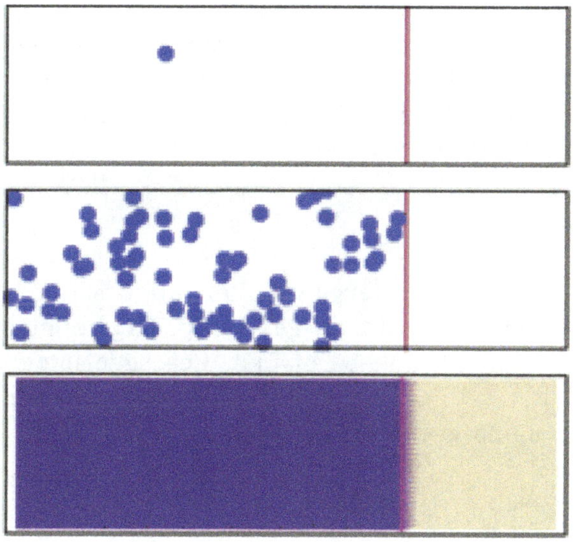

Diffusion is an example of the law of large numbers, applied to chemistry. Initially, there are solute molecules on the left side of a barrier (purple line) and none on the right. The barrier is removed, and the solute diffuses to fill the whole container.

Top: With a single molecule, the motion appears to be quite random.

Middle: With more molecules, there is clearly a trend where the solute fills the container more and more uniformly, but there are also random fluctuations.

Bottom: With an enormous number of solute molecules (too many to see), the randomness is essentially gone: The solute appears to move smoothly and systematically from high-concentration areas to low-concentration areas. In realistic situations, chemists can describe diffusion as a deterministic macroscopic phenomenon (see Fick`s laws), despite its underlying random nature.

The Italian mathematician Gerolamo Cardano (1501–1576) stated without proof that the accuracies of empirical statistics tend to improve with the number of trials. This was then formalized as a law of large numbers. A special form of the LLN (for a binary random variable) was first proved by Jacob Bernoulli. It took him over 20 years to develop a sufficiently rigorous mathematical proof which was published in his Ars Conjectandi (The Art of Conjecturing) in 1713. He named this his "Golden Theorem" but it became generally known as "**Bernoulli`s Theorem**". This should not be confused with Bernoulli's principle, named after Jacob Bernoulli's nephew Daniel Bernoulli. In 1837, S.D. Poisson further described it under the name "la loi des grands nombres" ("The law of large numbers").Thereafter, it was known under both names, but the "Law of large numbers" is most frequently used.

After Bernoulli and Poisson published their efforts, other mathematicians also contributed to refinement of the law, including Chebyshev, Markov, Borel, Cantelli and Kolmogorov and Khinchin. Markov showed that the law can apply to a random variable that does not have a finite variance under some other weaker assumption, and Khinchin showed in 1929 that if the series consists of independent identically distributed random variables, it suffices that the expected value exists for the weak law of large numbers to be true. These further studies have given rise to two prominent forms of the LLN. One is called the "weak" law and the other the "strong" law, in reference to two different modes of convergence of the cumulative sample means to the expected value; in particular, as explained below, the strong form implies the weak.

FORMS

Two different versions of the **law of large numbers** are described below; they are called the ***strong law*** of large numbers, and the ***weak law*** of large numbers. Stated for the case where X_1, X_2, ... is an infinite sequence of i.i.d. Lebesgue integrable random variables with expected value $E(X_1) = E(X_2) = ... = \mu$, both versions of the law state that – with virtual certainty – the sample average

$$\overline{X}_n = \frac{1}{n}(X_1 + \cdots + X_n)$$

converges to the expected value

$$\overline{X}_n \to \mu \quad \text{for} \quad n \to \infty,$$

(Lebesgue integrability of X_j means that the expected value $E(X_j)$ exists according to Lebesgue integration and is finite.)

An assumption of finite variance $\mathrm{Var}(X_1) = \mathrm{Var}(X_2) = ... = \sigma^2 < \infty$ is **not necessary**. Large or infinite variance will make the convergence slower, but the LLN holds anyway. This assumption is often used because it makes the proofs easier and shorter.

The difference between the strong and the weak version is concerned with the mode of convergence being asserted. For interpretation of these modes, see Convergence of random variables.

Weak Law

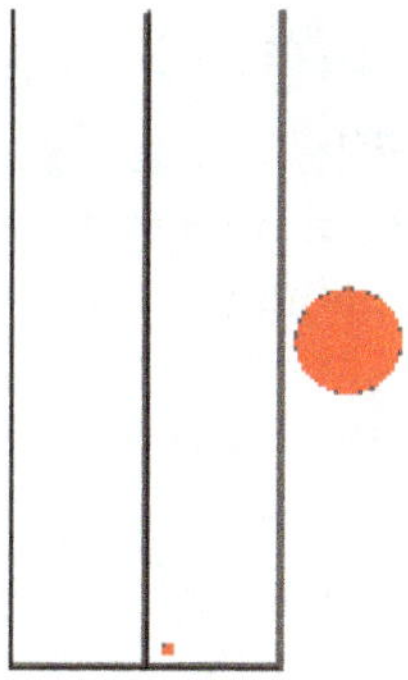

Simulation illustrating the law of large numbers. Each frame, you flip a coin that is red on one side and blue on the other, and put a dot in the corresponding column. A pie chart shows the proportion of red and blue so far. Notice that the proportion may vary a lot at first, but gradually approaches 50%.

The **weak law of large numbers** (also called Khintchine`s law) states that the sample average converges in probability towards the expected value

$$\overline{X}_n \xrightarrow{P} \mu \qquad \text{when } n \to \infty.$$

That is to say that for any positive number ε,

$$\lim_{n\to\infty} \Pr\Big(|\overline{X}_n - \mu| > \varepsilon \Big) = 0.$$

Interpreting this result, the weak law essentially states that for any nonzero margin specified, no matter how small, with a sufficiently large sample there will be a very high probability that the average of the observations will be close to the expected value; that is, within the margin.

Convergence in probability is also called weak convergence of random variables. This version is called the weak law because random variables may converge weakly (in probability) as above without converging strongly (almost surely) as below.

As mentioned earlier, the weak law applies in the case of independent identically distributed random variables having an expected value. But it also applies in some other cases. For example, the variance may be different for each random variable in the series, keeping the expected value constant. If the variances are bounded, then the law applies, as shown by Chebyshev as early as 1867. (If the expected values change during the series, then we can simply apply the law to the average deviation from the respective expected values. The law then states that this converges in probability to zero.) In fact, Chebyshev`s proof works so long as the variance of the average of the first n values goes to zero as n goes to infinity. As an example, assume that each random variable in the series follows a Gaussian distribution with mean zero, but with variance equal to $2n/\log(n+1)$ At each stage, the average will be normally distributed (since it is the average of a set of normally distributed variables). The variance of the sum is equal to the sum of the variances, which is asymptotic to $n^2/\log n$. The variance of the average is therefore asymptotic to $1/\log n$ and goes to zero.

An example where the law of large numbers does not apply is the Cauchy distribution. Let the random numbers equal the tangent of an angle uniformly distributed between −90° and +90°. The median is zero, but the expected value does not exist, and indeed the average of n such variables has the same distribution as one such variable. It does not tend toward zero as n goes to infinity.

There are also examples of the weak law applying even though the expected value does not exist. See #Differences between the weak law and the strong law.

Strong Law

The **strong law of large numbers** states that the sample average converges almost surely to the expected value

$$\bar{X}_n \xrightarrow{\text{a.s.}} \mu \qquad \text{when } n \to \infty.$$

That is,

$$\Pr\left(\lim_{n\to\infty} \bar{X}_n = \mu\right) = 1.$$

What this means is that as the number of trials n goes to infinity, the probability that the average of the observations is equal to the expected value will be equal to one.

The proof is more complex than that of the weak law. This law justifies the intuitive interpretation of the expected value (for Lebesgue integration only) of a random variable when sampled repeatedly as the long-term average".

Almost sure convergence is also called strong convergence of random variables. This version is called the strong law because random variables which converge strongly (almost surely) are guaranteed to converge weakly (in probability). The strong law implies the weak law but not vice versa, when the strong law conditions hold the variable converges both strongly (almost surely) and weakly (in probability). However the weak law may hold in conditions where the strong law does not hold and then the convergence is only weak (in probability).

To date it has not been possible to prove that the strong law conditions are the same as those of the weak law.

The strong law of large numbers can itself be seen as a special case of the pointwise ergodic theorem.

The strong law applies to independent identically distributed random variables having an expected value (like the weak law). This was proved by Kolmogorov in 1930. It can also apply in other cases. Kolmogorov also showed, in 1933, that if the variables are independent and identically distributed, then for the average to converge almost surely on something (this can be considered another statement of the strong law), it is necessary that they have an expected value (and then of course the average will converge almost surely on that).

If the summands are independent but not identically distributed, then

$$\bar{X}_n - \mathrm{E}\left[\bar{X}_n\right] \xrightarrow{\text{a.s.}} 0,$$

provided that each X_k has a finite second moment and

$$\sum_{k=1}^{\infty} \frac{1}{k^2} \mathrm{Var}[X_k] < \infty.$$

This statement is known as Kolmogorov's strong law, see e.g. Sen & Singer (1993, Theorem 2.3.10).

An example of a series where the weak law applies but not the strong law is when X_k is plus or minus $\sqrt{k/\log\log\log k}$ (starting at sufficiently large k so that the denominator is positive) with probability 1/2 for each. The variance of X_k is then $k/\log\log\log k$. Kolmogorov's strong law does not apply because the partial sum in his criterion up to k=n is asymptotic to $\log n/\log\log\log n$ and this is unbounded.

If we replace the random variables with Gaussian variables having the same variances, namely $\sqrt{k/\log\log\log k}$, then the average at any point will also be normally distributed. The width of the distribution of the average will tend toward zero (standard deviation asymptotic to $1/\sqrt{2\log\log\log n}$), but for a given ε, there is probability which does not go to zero with n that the average sometime after the nth trial will come back up to ε. Since this probability does not go to zero, it must have a positive lower bound p(ε), which means there is a probability of at least p(ε) that the average will attain ε after n trials. It will happen with probability p(ε)/2 before some m which depends on n. But even after m, there is still a probability of at least p(ε) that it will happen. (This seems to indicate that p(ε)=1 and the average will attain ε an infinite number of times.)

Differences between the Weak Law and the Strong Law

The weak law states that for a specified large n, the average $\overline{X}_n$ is likely to be near μ. Thus, it leaves open the possibility that $|\overline{X}_n - \mu| > \varepsilon$ happens an infinite number of times, although at infrequent intervals. (Not necessarily $|\overline{X}_n - \mu| \neq 0$ for all n).

The strong law shows that this almost surely will not occur. In particular, it implies that with probability 1, we have that for any $\varepsilon > 0$ the inequality $|\overline{X}_n - \mu| < \varepsilon$ holds for all large enough n.

The strong law does not hold in the following cases, but the weak law does.

1. Let x be an exponentially distributed random variable with parameter

1. The random variable $\frac{\sin(x)e^x}{x}$ has no expected value according to Lebesgue integration, but using conditional convergence and interpreting the integral as a Dirichlet integral, which is an improper Riemann integral, we can say:

$$E\left(\frac{\sin(x)e^x}{x}\right) = \int_0^{\infty} \frac{\sin(x)e^x}{x} e^{-x}\, dx = \frac{\pi}{2}$$

2. Let x be geometric distribution with probability 0.5. The random variable $\frac{2^x(-1)^x}{x}$ does not have an expected value in the conventional sense because the infinite series is not absolutely convergent, but using conditional convergence, we can say:

$$E\left(\frac{2^x(-1)^x}{x}\right) = \sum_{1}^{\infty} \frac{2^x(-1)^x}{x} 2^{-x} = -\ln(2)$$

3. If the cumulative distribution function of a random variable is

$$1 - F(x) = \frac{e}{2x\ln(x)}, x \geq e$$

$$F(x) = \frac{e}{-2x\ln(-x)}, x \leq -e$$

then it has no expected value, but the weak law is true.

Uniform law of Large Numbers

Suppose f(x,θ) is some function defined for θ ∈ Θ, and continuous in θ. Then for any fixed θ, the sequence $\{f(X_1,\theta), f(X_2,\theta), \ldots\}$ will be a sequence of independent and identically distributed random variables, such that the sample mean of this sequence converges in probability to E[f(X,θ)]. This is the pointwise (in θ) convergence.

The **uniform law of large numbers** states the conditions under which the convergence happens uniformly in θ. If

- Θ is compact,
- f(x,θ) is continuous at each θ ∈ Θ for almost all x's, and measurable function of x at each θ.
- there exists a dominating function d(x) such that E[d(X)] < ∞, and

$$\|f(x,\theta)\| \leq d(x) \quad \text{for all } \theta \in \Theta.$$

Then E[f(X,θ)] is continuous in θ, and

$$\sup_{\theta\in\Theta}\left\|\frac{1}{n}\sum_{i=1}^{n}f(X_i,\theta)-\mathrm{E}[f(X,\theta)]\right\|\xrightarrow{\text{a.s.}}0.$$

This result is useful to derive consistency of a large class of estimators (see Extremum estimator).

Borel's law of Large Numbers

Borel's law of large numbers, named after Émile Borel, states that if an experiment is repeated a large number of times, independently under identical conditions, then the proportion of times that any specified event occurs approximately equals the probability of the event`s occurrence on any particular trial; the larger the number of repetitions, the better the approximation tends to be. More precisely, if E denotes the event in question, p its probability of occurrence, and $N_n(E)$ the number of times E occurs in the first n trials, then with probability one,

$$\frac{N_n(E)}{n}\to p \text{ as } n\to\infty.$$

Chebyshev's inequality. Let X be a random variable with finite expected value μ and finite non-zero variance σ^2. Then for any real number k > 0,

$$\Pr(|X-\mu|\geq k\sigma)\leq\frac{1}{k^2}$$

This theorem makes rigorous the intuitive notion of probability as the long-run relative frequency of an event's occurrence. It is a special case of any of several more general laws of large numbers in probability theory.

Theorem: $\overline{X}_n \xrightarrow{P} \mu \qquad \text{when } n\to\infty.$

Proof of the Weak Law

Given X_1, X_2, ... an infinite sequence of i.i.d. random variables with finite expected value $E(X_1) = E(X_2) = \ldots = \mu < \infty$, we are interested in the convergence of the sample average $\overline{X}_n=\tfrac{1}{n}(X_1+\cdots+X_n).$

The weak law of large numbers states:

Proof using Chebyshev's inequality

This proof uses the assumption of finite variance : $\operatorname{Var}(X_i)=\sigma^2$ (for all i). The independence of the random variables implies no correlation between them, and we have that

$$\operatorname{Var}(\overline{X}_n) = \operatorname{Var}(\tfrac{1}{n}(X_1 + \cdots + X_n)) = \frac{1}{n^2}\operatorname{Var}(X_1 + \cdots + X_n) = \frac{n\sigma^2}{n^2} = \frac{\sigma^2}{n}.$$

The common mean μ of the sequence is the mean of the sample average:

$$E(\overline{X}_n) = \mu.$$

Using Chebyshev's inequality on $\overline{X}_n$ results in

$$\mathrm{P}(|\overline{X}_n - \mu| \geq \varepsilon) \leq \frac{\sigma^2}{n\varepsilon^2}.$$

This may be used to obtain the following:

$$\mathrm{P}(|\overline{X}_n - \mu| < \varepsilon) = 1 - \mathrm{P}(|\overline{X}_n - \mu| \geq \varepsilon) \geq 1 - \frac{\sigma^2}{n\varepsilon^2}.$$

As n approaches infinity, the expression approaches 1. And by definition of convergence in probability, we have obtained

$$\overline{X}_n \xrightarrow{P} \mu \qquad \text{when } n \to \infty.$$

Proof using Convergence of Characteristic Functions

By Taylor's theorem for complex functions, the characteristic function of any random variable, X, with finite mean μ, can be written as

$$\varphi_X(t) = 1 + it\mu + o(t), \quad t \to 0.$$

All $X_1, X_2, \ldots$ have the same characteristic function, so we will simply denote this φ_X.

Among the basic properties of characteristic functions there are

$$\varphi_{\frac{1}{n}X}(t) = \varphi_X(\tfrac{t}{n}) \quad \text{and} \quad \varphi_{X+Y}(t) = \varphi_X(t)\varphi_Y(t)$$ if X and Y are independent.

These rules can be used to calculate the characteristic function of $\overline{X}_n$ in terms of φ_X:

$$\varphi_{\overline{X}_n}(t) = \left[\varphi_X\left(\frac{t}{n}\right)\right]^n = \left[1 + i\mu\frac{t}{n} + o\left(\frac{t}{n}\right)\right]^n \to e^{it\mu}, \quad \text{as} \quad n \to \infty.$$

The limit $e^{it\mu}$ is the characteristic function of the constant random variable μ, and hence by the Lévy continuity theorem, $\overline{X}_n$ converges in distribution to μ:

$$\overline{X}_n \xrightarrow{\mathcal{D}} \mu \qquad \text{for} \qquad n \to \infty.$$

μ is a constant, which implies that convergence in distribution to μ and convergence in probability to μ are equivalent (see Convergence of random variables.) Therefore,

$$\overline{X}_n \xrightarrow{P} \mu \qquad \text{when } n \to \infty.$$

This shows that the sample mean converges in probability to the derivative of the characteristic function at the origin, as long as the latter exists.

CENTRAL LIMIT THEOREM

In probability theory, the **central limit theorem** (**CLT**) establishes that, for the most commonly studied scenarios, when independent random variables are added, their sum tends toward a normal distribution (commonly known as a bell curve) even if the original variables themselves are not normally distributed. In more precise terms, given certain conditions, the arithmetic mean of a sufficiently large number of iterates of independent random variables, each with a well-defined (finite) expected value and finite variance, will be approximately normally distributed, regardless of the underlying distribution. The theorem is a key concept in probability theory because it implies that probabilistic and statistical methods that work for normal distributions can be applicable to many problems involving other types of distributions.

To illustrate the meaning of the theorem, suppose that a sample is obtained containing a large number of observations, each observation being randomly generated in a way that does not depend on the values of the other observations, and that the arithmetic average of the observed values is computed. If this procedure is performed many times, the central limit theorem says that the computed values of the average will be distributed according to the normal distribution (commonly known as a "bell curve"). A simple example of this is that if one flips a coin many times the probability of getting a given number of heads in a series of flips should follow a normal curve, with mean equal to half the total number of flips in each series.

The central limit theorem has a number of variants. In its common form, the random variables must be identically distributed. In variants, convergence of the mean to the normal distribution also occurs for non-identical distributions or for non-independent observations, given that they comply with certain conditions.

In more general usage, a **central limit theorem** is any of a set of weak-convergence theorems in probability theory. They all express the fact that a sum of many independent and identically distributed (i.i.d.) random variables, or alternatively, random variables with specific types of dependence, will tend to be distributed according to one of a small set of attractor distributions. When the variance of the i.i.d. variables is finite, the attractor distribution is the normal distribution. In contrast, the sum of a number of i.i.d. random variables with power law tail distributions decreasing as $|x|^{-\alpha-1}$ where $0 < \alpha < 2$ (and therefore having infinite variance) will tend to an alpha-stable distribution with stability parameter (or index of stability) of α as the number of variables grows.

Central limit Theorems for Independent Sequences

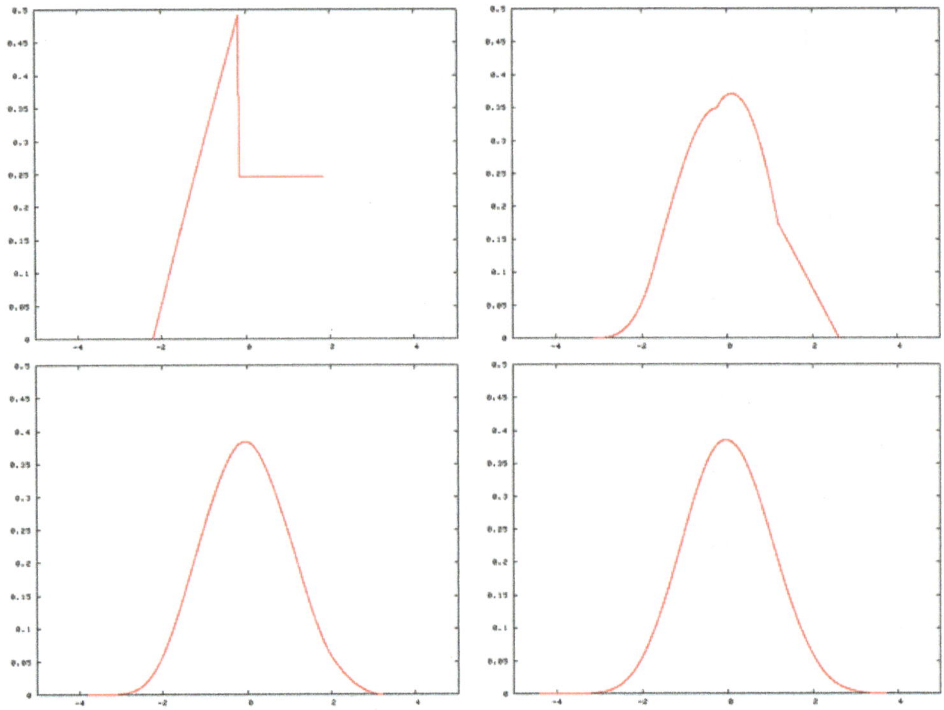

A distribution being "smoothed out" by summation, showing original density of distribution and three subsequent summations; see Illustration of the central limit theorem for further details.

Classical CLT

Let $\{X_1, ..., X_n\}$ be a random sample of size n — that is, a sequence of independent and identically distributed random variables drawn from distributions of expected values given by μ and finite variances given by σ^2. Suppose we are interested in the sample average

$$S_n := \frac{X_1 + \cdots + X_n}{n}$$

of these random variables. By the law of large numbers, the sample averages converge in probability and almost surely to the expected value μ as $n \to \infty$. The classical central limit theorem describes the size and the distributional form of the stochastic fluctuations around the deterministic number μ during this convergence. More precisely, it states that as n gets larger, the distribution of the difference between the sample average S_n and its limit μ, when multiplied by the factor $\sqrt{n}$ (that is $\sqrt{n}(S_n - \mu)$), approximates the normal distribution with mean 0 and variance σ^2. For large enough n, the distribution of S_n is close to the normal distribution with mean μ and variance σ^2/n. The usefulness of the theorem is that the distribution of $\sqrt{n}(S_n - \mu)$ approaches normality regardless of the shape of the distribution of the individual X_i's. Formally, the theorem can be stated as follows:

Lindeberg–Lévy CLT. Suppose $\{X_1, X_2, ...\}$ is a sequence of i.i.d. random variables with $E[X_i] = \mu$ and $Var[X_i] = \sigma^2 < \infty$. Then as n approaches infinity, the random variables $\sqrt{n}(S_n - \mu)$ converge in distribution to a normal $N(0, \sigma^2)$:

$$\sqrt{n}\left(\left(\frac{1}{n}\sum_{i=1}^{n} X_i\right) - \mu\right) \xrightarrow{d} N(0, \sigma^2).$$

In the case $\sigma > 0$, convergence in distribution means that the cumulative distribution functions of $\sqrt{n}(S_n - \mu)$ converge pointwise to the cdf of the $N(0, \sigma^2)$ distribution: for every real number z,

$$\lim_{n\to\infty} \Pr[\sqrt{n}(S_n - \mu) \le z] = \Phi(z/\sigma),$$

where $\Phi(x)$ is the standard normal cdf evaluated at x. Note that the convergence is uniform in z in the sense that

$$\lim_{n\to\infty} \sup_{z\in\mathbf{R}} \left|\Pr[\sqrt{n}(S_n - \mu) \le z] - \Phi(z/\sigma)\right| = 0,$$

where sup denotes the least upper bound (or supremum) of the set.

Lyapunov CLT

The theorem is named after Russian mathematician Aleksandr Lyapunov. In this variant of the central limit theorem the random variables X_i have to be independent, but not necessarily identically distributed. The theorem also requires that random variables $|X_i|$ have moments of some order $(2 + \delta)$, and that the rate of growth of these moments is limited by the Lyapunov condition given below.

Lyapunov CLT. Suppose $\{X_1, X_2, ...\}$ is a sequence of independent random variables, each with finite expected value μ_i and variance σ 2i . Define

$$s_n^2 = \sum_{i=1}^{n} \sigma_i^2$$

If for some $\delta > 0$, the *Lyapunov's condition*

$$\lim_{n\to\infty} \frac{1}{s_n^{2+\delta}} \sum_{i=1}^{n} \mathrm{E}\left[|X_i - \mu_i|^{2+\delta}\right] = 0$$

is satisfied, then a sum of $(X_i - \mu_i)/s_n$ converges in distribution to a standard normal random variable, as *n* goes to infinity:

$$\frac{1}{s_n}\sum_{i=1}^{n}(X_i - \mu_i) \xrightarrow{d} \mathcal{N}(0, 1).$$

In practice it is usually easiest to check the Lyapunov's condition for $\delta = 1$. If a sequence of random variables satisfies Lyapunov's condition, then it also satisfies Lindeberg's condition. The converse implication, however, does not hold.

Lindeberg CLT

In the same setting and with the same notation as above, the Lyapunov condition can be replaced with the following weaker one (from Lindeberg in 1920).

Suppose that for every ε > 0

$$\lim_{n\to\infty} \frac{1}{s_n^2}\sum_{i=1}^{n} \mathbf{E}\left[(X_i-\mu_i)^2\cdot \mathbf{1}_{\{|X_i-\mu_i|>\varepsilon s_n\}}\right]=0$$

where $\mathbf{1}_{\{\ldots\}}$ is the indicator function. Then the distribution of the standardized sums $\frac{1}{s_n}\sum_{i=1}^{n}(X_i-\mu_i)$ converges towards the standard normal distribution N(0,1).

Multidimensional CLT

Proofs that use characteristic functions can be extended to cases where each individual X_i is a random vector in $\mathbf{R}^k$, with mean vector $\mu = E(X_i)$ and covariance matrix Σ (amongst the components of the vector), and these random vectors are independent and identically distributed. Summation of these vectors is being done componentwise. The multidimensional central limit theorem states that when scaled, sums converge to a multivariate normal distribution.

Let

$$\mathbf{X_i}=\begin{bmatrix}X_{i(1)}\\ \vdots\\ X_{i(k)}\end{bmatrix}$$

be the k-vector. The bold in $\mathbf{X}_i$ means that it is a random vector, not a random (univariate) variable. Then the sum of the random vectors will be

$$\begin{bmatrix}X_{1(1)}\\ \vdots\\ X_{1(k)}\end{bmatrix}+\begin{bmatrix}X_{2(1)}\\ \vdots\\ X_{2(k)}\end{bmatrix}+\cdots+\begin{bmatrix}X_{n(1)}\\ \vdots\\ X_{n(k)}\end{bmatrix}=\begin{bmatrix}\sum_{i=1}^{n}\left[X_{i(1)}\right]\\ \vdots\\ \sum_{i=1}^{n}\left[X_{i(k)}\right]\end{bmatrix}=\sum_{i=1}^{n}\mathbf{X_i}$$

and the average is

$$\frac{1}{n}\sum_{i=1}^{n}\mathbf{X_i}=\frac{1}{n}\begin{bmatrix}\sum_{i=1}^{n}X_{i(1)}\\ \vdots\\ \sum_{i=1}^{n}X_{i(k)}\end{bmatrix}=\begin{bmatrix}\bar{X}_{i(1)}\\ \vdots\\ \bar{X}_{i(k)}\end{bmatrix}=\mathbf{\bar{X}_n}$$

and therefore

$$\frac{1}{\sqrt{n}}\sum_{i=1}^{n}\left[\mathbf{X_i}-E\left(X_i\right)\right]=\frac{1}{\sqrt{n}}\sum_{i=1}^{n}(\mathbf{X_i}-\mu)=\sqrt{n}\left(\overline{\mathbf{X}}_n-\mu\right).$$

The multivariate central limit theorem states that

$$\sqrt{n}\left(\overline{\mathbf{X}}_n-\mu\right)\xrightarrow{D}\mathcal{N}_k(0,\Sigma)$$

where the covariance matrix Σ is equal to

$$\Sigma = \begin{bmatrix} \text{Var}(X_{1(1)}) & \text{Cov}(X_{1(1)}, X_{1(2)}) & \text{Cov}(X_{1(1)}, X_{1(3)}) & \cdots & \text{Cov}(X_{1(1)}, X_{1(k)}) \\ \text{Cov}(X_{1(2)}, X_{1(1)}) & \text{Var}(X_{1(2)}) & \text{Cov}(X_{1(2)}, X_{1(3)}) & \cdots & \text{Cov}(X_{1(2)}, X_{1(k)}) \\ \text{Cov}(X_{1(3)}, X_{1(1)}) & \text{Cov}(X_{1(3)}, X_{1(2)}) & \text{Var}(X_{1(3)}) & \cdots & \text{Cov}(X_{1(3)}, X_{1(k)}) \\ \vdots & \vdots & \vdots & \ddots & \vdots \\ \text{Cov}(X_{1(k)}, X_{1(1)}) & \text{Cov}(X_{1(k)}, X_{1(2)}) & \text{Cov}(X_{1(k)}, X_{1(3)}) & \cdots & \text{Var}(X_{1(k)}) \end{bmatrix}$$

Generalised Theorem

The central limit theorem states that the sum of a number of independent and identically distributed random variables with finite variances will tend to a normal distribution as the number of variables grows. A generalization due to Gnedenko and Kolmogorov states that the sum of a number of random variables with a power-law tail (Paretian tail) distributions decreasing as $|x|^{-\alpha-1}$ where $0 < \alpha < 2$ (and therefore having infinite variance) will tend to a stable distribution $f(x; \alpha, 0, c, 0)$ as the number of summands grows. If $\alpha>2$ then the sum converges to a stable distribution with stability parameter equal to 2, i.e. a Gaussian distribution.

Central Limit Theorems for Dependent Processes

CLT under Weak Dependence

A useful generalization of a sequence of independent, identically distributed random variables is a mixing random process in discrete time; "mixing" means, roughly, that random variables temporally far apart from one another are nearly independent. Several kinds of mixing are used in ergodic theory and probability theory. See especially strong mixing (also called α-mixing) defined by $\alpha(n) \to 0$ where $\alpha(n)$ is so-called strong mixing coefficient.

A simplified formulation of the central limit theorem under strong mixing is:

Theorem. Suppose that X_1, X_2, ... is stationary and α-mixing with $\alpha_n = O(n^{-5})$ and that $E(X_n) = 0$ and $E(X_n^{12}) < \infty$. Denote $S_n = X_1 + ... + X_n$, then the limit

$$\sigma^2 = \lim_n \frac{E(S_n^2)}{n}$$

exists, and if $\sigma \neq 0$ then $S_n/(\sigma\sqrt{n})$ converges in distribution to N(0, 1).

In fact,

$$\sigma^2 = E(X_1^2) + 2\sum_{k=1}^{\infty} E(X_1 X_{1+k}),$$

where the series converges absolutely.

The assumption $\sigma \neq 0$ cannot be omitted, since the asymptotic normality fails for $X_n = Y_n - Y_{n-1}$ where Y_n are another stationary sequence.

There is a stronger version of the theorem: the assumption $E(X_n^{12}) < \infty$ is replaced with $E(|X_n|^{2+\delta}) < \infty$, and the assumption $\alpha_n = O(n^{-5})$ is replaced with {\displaystyle \

sum _{n}\alpha _{n}^{\frac {\delta }{2(2+\delta)}}<\infty .} Existence of such $\delta > 0$ ensures the conclusion. For encyclopedic treatment of limit theorems under mixing conditions see (Bradley 2007).

Martingale difference CLT

Main article: Martingale central limit theorem

Theorem. Let a martingale M_n satisfy

- $\frac{1}{n}\sum_{k=1}^{n} \mathrm{E}((M_k - M_{k-1})^2 | M_1, \dots, M_{k-1}) \to 1$ in probability as n tends to infinity,
- for every $\varepsilon > 0$, $\frac{1}{n}\sum_{k=1}^{n} \mathrm{E}\left((M_k - M_{k-1})^2; |M_k - M_{k-1}| > \varepsilon\sqrt{n}\right) \to 0$ as n tends to infinity,

then $M_n/\sqrt{n}$ converges in distribution to $N(0,1)$ as $n \to \infty$.

Caution: The restricted expectation $E(X; A)$ should not be confused with the conditional expectation $E(X \mid A) = E(X; A)/\mathbf{P}(A)$.

Remarks

Proof of classical CLT

For a theorem of such fundamental importance to statistics and applied probability, the central limit theorem has a remarkably simple proof using characteristic functions. It is similar to the proof of the (weak) law of large numbers.

As stated above, suppose $\{X_1, \dots, X_n\}$ are independent and identically distributed random variables, each with mean μ and finite variance σ^2. The sum $X_1+\dots+X_n$ has mean $n\mu$ and variance $n\sigma^2$. Consider the random variable

$$Z_n = \frac{X_1 + \cdots + X_n - n\mu}{\sqrt{n\sigma^2}} = \sum_{i=1}^{n} \frac{X_i - \mu}{\sqrt{n\sigma^2}} = \sum_{i=1}^{n} \frac{1}{\sqrt{n}} Y_i,$$

where in the last step we defined the new random variables $Y_i = (X_i - \mu)/\sigma$, each with zero mean and unit variance ($\mathrm{var}(Y) = 1$). The characteristic function of Z_n is given by

$$\varphi_{Z_n} = \varphi_{\sum_{i=1}^{n} \frac{1}{\sqrt{n}} Y_i}(t) = \varphi_{Y_1}(t/\sqrt{n}) \cdot \varphi_{Y_2}(t/\sqrt{n}) \cdots \varphi_{Y_n}(t/\sqrt{n})$$

$$= \left[\varphi_{Y_1}\left(\frac{t}{\sqrt{n}}\right)\right]^n,$$

where in the last step we used the fact that all of the Y_i are identically distributed. The characteristic function of Y_1 is, by Taylor`s theorem,

$$\varphi_{Y_1}\left(\frac{t}{\sqrt{n}}\right) = 1 - \frac{t^2}{2n} + c\frac{t^3}{6n^{3/2}} + o\left(\frac{t^3}{n^{3/2}}\right), \quad t \to 0$$

where c is a (complex) constant and $o(t^3)$ is "little o notation" for some function of t that goes to zero more rapidly than t^3. By the limit of the exponential function ($e^x = \lim(1 + x/n)^n$), the characteristic function of Z_n equals

$$\varphi_{Z_n} = \left(1 - \frac{t^2}{2n} + c\frac{t^3}{6n^{3/2}} + o\left(\frac{t^3}{n^{3/2}}\right)\right)^n \to e^{-t^2/2}, \quad n \to \infty.$$

Note that all of the higher order terms vanish in the limit $n \to \infty$. The right hand side equals the characteristic function of a standard normal distribution N(0, 1), which implies through Lévy's continuity theorem that the distribution of Z_n will approach N(0, 1) as $n \to \infty$. Therefore, the sum $X_1 + \ldots + X_n$ will approach that of the normal distribution $N(n\mu, n\sigma^2)$, and the sample average

$$S_n = \frac{X_1 + \cdots + X_n}{n}$$

converges to the normal distribution $N(\mu, \sigma^2/n)$, from which the central limit theorem follows.

Convergence to the Limit

The central limit theorem gives only an asymptotic distribution. As an approximation for a finite number of observations, it provides a reasonable approximation only when close to the peak of the normal distribution; it requires a very large number of observations to stretch into the tails.

The convergence in the Central limit theorem is uniform because the limiting cumulative distribution function is continuous. If the third central moment $E((X_1 - \mu)^3)$ exists and is finite, then the speed of convergence is at least on the order of $1/n^{1/2}$(see Berry-Esseen theorem). Stein's method can be used not only to prove the central limit theorem, but also to provide bounds on the rates of convergence for selected metrics.

The convergence to the normal distribution is monotonic, in the sense that the entropy of Z_n increases monotonically to that of the normal distribution.

The central limit theorem applies in particular to sums of independent and identically distributed discrete random variables. A sum of discrete random variables is still a discrete random variable, so that we are confronted with a sequence of discrete random variables whose cumulative probability distribution function converges towards a cumulative probability distribution function corresponding to a continuous variable (namely that of the normal distribution). This means that if we build a histogram of the realisations of the sum of n independent identical discrete variables, the curve that joins the centers of the upper faces of the rectangles forming the histogram converges toward a Gaussian curve as n approaches infinity, this relation is known as de Moivre–Laplace theorem. The binomial distribution article details such an application of the central limit theorem in the simple case of a discrete variable taking only two possible values.

Relation to the Law of Large Numbers

The law of large numbers as well as the central limit theorem are partial solutions to a general problem: "What is the limiting behaviour of S_n as n approaches infinity?" In mathematical analysis, asymptotic series are one of the most popular tools employed to approach such questions.

Suppose we have an asymptotic expansion of f(n):

$$f(n) = a_1\varphi_1(n) + a_2\varphi_2(n) + O(\varphi_3(n)) \qquad (n \to \infty).$$

Dividing both parts by $\varphi_1(n)$ and taking the limit will produce a_1, the coefficient of the highest-order term in the expansion, which represents the rate at which f(n) changes in its leading term.

$$\lim_{n\to\infty} \frac{f(n)}{\varphi_1(n)} = a_1.$$

Informally, one can say: "f(n) grows approximately as $a_1\ \varphi_1(n)$". Taking the difference between f(n) and its approximation and then dividing by the next term in the expansion, we arrive at a more refined statement about f(n):

$$\lim_{n\to\infty} \frac{f(n) - a_1\varphi_1(n)}{\varphi_2(n)} = a_2.$$

Here one can say that the difference between the function and its approximation grows approximately as $a_2\ \varphi_2(n)$. The idea is that dividing the function by appropriate normalizing functions, and looking at the limiting behavior of the result, can tell us much about the limiting behavior of the original function itself.

Informally, something along these lines is happening when the sum, S_n, of independent identically distributed random variables, X_1, ..., X_n, is studied in classical probability theory. If each X_i has finite mean μ, then by the law of large numbers, $S_n/n \to \mu$. If in addition each X_i has finite variance σ^2, then by the central limit theorem,

$$\frac{S_n - n\mu}{\sqrt{n}} \to \xi,$$

where ξ is distributed as $N(0, \sigma^2)$. This provides values of the first two constants in the informal expansion

$$S_n \approx \mu n + \xi\sqrt{n}.$$

In the case where the X_i's do not have finite mean or variance, convergence of the shifted and rescaled sum can also occur with different centering and scaling factors:

$$\frac{S_n - a_n}{b_n} \to \Xi,$$

or informally

$$S_n \approx a_n + \Xi b_n.$$

Distributions Ξ which can arise in this way are called stable. Clearly, the normal distribution is stable, but there are also other stable distributions, such as the Cauchy distribution, for which the mean or variance are not defined. The scaling factor b_n may be proportional to n^c, for any $c \geq 1/2$; it may also be multiplied by a slowly varying function of n.

The law of the iterated logarithm specifies what is happening "in between" the law of large numbers and the central limit theorem. Specifically it says that the normalizing function $\sqrt{n \log \log n}$ intermediate in size between n of the law of large numbers and $\sqrt{n}$ of the central limit theorem provides a non-trivial limiting behavior.

Alternative statements of the theorem

Density Functions

The density of the sum of two or more independent variables is the convolution of their densities (if these densities exist). Thus the central limit theorem can be interpreted as a statement about the properties of density functions under convolution: the convolution of a number of density functions tends to the normal density as the number of density functions increases without bound. These theorems require stronger hypotheses than the forms of the central limit theorem given above. Theorems of this type are often called local limit theorems. See Petrov for a particular local limit theorem for sums of independent and identically distributed random variables.

Characteristic Functions

Since the characteristic function of a convolution is the product of the characteristic functions of the densities involved, the central limit theorem has yet another restatement: the product of the characteristic functions of a number of density functions becomes close to the characteristic function of the normal density as the number of density functions increases without bound, under the conditions stated above. However, to state this more precisely, an appropriate scaling factor needs to be applied to the argument of the characteristic function.

An equivalent statement can be made about Fourier transforms, since the characteristic function is essentially a Fourier transform.

Extensions to the Theorem

Products of Positive Random Variables

The logarithm of a product is simply the sum of the logarithms of the factors. Therefore, when the logarithm of a product of random variables that take only positive values approaches a normal distribution, the product itself approaches a log-normal distribution. Many physical quantities (especially mass or length, which are a matter of scale and cannot be negative) are the products of different random factors, so they follow a log-normal distribution. This multiplicative version of the central limit theorem is sometimes called Gibrat`s law.

Whereas the central limit theorem for sums of random variables requires the condition of finite variance, the corresponding theorem for products requires the corresponding condition that the density function be square-integrable.

Beyond the Classical Framework

Asymptotic normality, that is, convergence to the normal distribution after appropriate shift and rescaling, is a phenomenon much more general than the classical framework treated above, namely, sums of independent random variables (or vectors). New frameworks are revealed from time to time; no single unifying framework is available for now.

Convex Body

Theorem. There exists a sequence $\varepsilon_n \downarrow 0$ for which the following holds. Let $n \geq 1$, and let random variables $X_1, \ldots, X_n$ have a log-concave joint density f such that $f(x_1, \ldots, x_n) = f(|x_1|, \ldots, |x_n|)$ for all $x_1, \ldots, x_n$, and $E(X_k^2) = 1$ for all $k = 1, \ldots, n$. Then the distribution of

$$\frac{X_1 + \cdots + X_n}{\sqrt{n}}$$

is ε_n-close to N(0, 1) in the total variation distance.

These two ε_n-close distributions have densities (in fact, log-concave densities), thus, the total variance distance between them is the integral of the absolute value of the difference between the densities. Convergence in total variation is stronger than weak convergence.

An important example of a log-concave density is a function constant inside a given convex body and vanishing outside; it corresponds to the uniform distribution on the convex body, which explains the term "central limit theorem for convex bodies".

Another example: $f(x_1, \ldots, x_n) = \text{const} \cdot \exp(-(|x_1|^\alpha + \ldots + |x_n|^\alpha)^\beta)$ where $\alpha > 1$ and $\alpha\beta > 1$. If $\beta = 1$ then $f(x_1, \ldots, x_n)$ factorizes into $\text{const} \cdot \exp(-|x_1|^\alpha)\ldots\exp(-|x_n|^\alpha)$, which means independence of $X_1, \ldots, X_n$. In general, however, they are dependent.

The condition $f(x_1, \ldots, x_n) = f(|x_1|, \ldots, |x_n|)$ ensures that $X_1, \ldots, X_n$ are of zero mean and uncorrelated; still, they need not be independent, nor even pairwise independent. By the way, pairwise independence cannot replace independence in the classical central limit theorem.

Here is a Berry–Esseen type result.

Theorem. Let $X_1, \ldots, X_n$ satisfy the assumptions of the previous theorem, then

$$\left|\mathbb{P}\left(a \leq \frac{X_1 + \cdots + X_n}{\sqrt{n}} \leq b\right) - \frac{1}{\sqrt{2\pi}} \int_a^b e^{-t^2/2}\, dt\right| \leq \frac{C}{n}$$

for all $a < b$; here C is a universal (absolute) constant. Moreover, for every $c_1, \ldots, c_n \in \mathbf{R}$ such that $c_1^2 + \ldots + c_n^2 = 1$,

$$\left|\mathbb{P}(a \leq c_1X_1 + \cdots + c_nX_n \leq b) - \frac{1}{\sqrt{2\pi}}\int_a^b e^{-t^2/2}\,dt\right| \leq C(c_1^4 + \cdots + c_n^4).$$

The distribution of $(X_1 + \cdots + X_n)/\sqrt{n}$ need not be approximately normal (in fact, it can be uniform). However, the distribution of $c_1X_1 + \ldots + c_nX_n$ is close to N(0, 1) (in the total variation distance) for most of vectors $(c_1, \ldots, c_n)$ according to the uniform distribution on the sphere $c_1^2 + \ldots + c_n^2 = 1$.

Lacunary Trigonometric Series

Theorem (Salem–Zygmund). Let U be a random variable distributed uniformly on $(0, 2\pi)$, and $X_k = r_k \cos(n_kU + a_k)$, where

- n_k satisfy the lacunarity condition: there exists $q > 1$ such that $n_{k+1} \geq qn_k$ for all k,
- r_k are such that

$$r_1^2 + r_2^2 + \cdots = \infty \text{ and } \frac{r_k^2}{r_1^2 + \cdots + r_k^2} \to 0,$$

- $0 \leq a_k < 2\pi$.

Then

$$\frac{X_1 + \cdots + X_k}{\sqrt{r_1^2 + \cdots + r_k^2}}$$

converges in distribution to N(0, 1/2).

Gaussian Polytopes

Theorem Let $A_1, \ldots, A_n$ be independent random points on the plane $\mathbf{R}^2$each having the two-dimensional standard normal distribution. Let K_n be the convex hull of these points, and X_n the area of K_n Then

$$\frac{X_n - \mathrm{E}(X_n)}{\sqrt{\mathrm{Var}(X_n)}}$$

converges in distribution to N(0, 1) as n tends to infinity.

The same holds in all dimensions (2, 3, ...).

The polytope K_n is called Gaussian random polytope.

A similar result holds for the number of vertices (of the Gaussian polytope), the number of edges, and in fact, faces of all dimensions.

Linear Functions of Orthogonal Matrices

A linear function of a matrix M is a linear combination of its elements (with given coefficients), $M \mapsto tr(AM)$ where A is the matrix of the coefficients; see Trace (linear algebra)#Inner product.

A random orthogonal matrix is said to be distributed uniformly, if its distribution is the normalized Haar measure on the orthogonal group $O(n, \mathbf{R})$; see Rotation matrix#Uniform random rotation matrices.

Theorem. Let M be a random orthogonal $n \times n$ matrix distributed uniformly, and A a fixed $n \times n$ matrix such that $tr(AA^*) = n$, and let $X = tr(AM)$. Then the distribution of X is close to $N(0, 1)$ in the total variation metric up to $2\sqrt{3}/(n - 1)$.

Subsequences

Theorem. Let random variables $X_1, X_2, \ldots \in L_2(\Omega)$ be such that $X_n \to 0$ weakly in $L_2(\Omega)$ and $X_n^2 \to 1$ weakly in $L_1(\Omega)$. Then there exist integers $n_1 < n_2 < \ldots$ such that $(X_{n_1} + \cdots + X_{n_k})/\sqrt{k}$ converges in distribution to $N(0, 1)$ as k tends to infinity.

Random walk on a Crystal Lattice

The central limit theorem may be established for the simple random walk on a crystal lattice (an infinite-fold abelian covering graph over a finite graph), and is used for design of crystal structures.

Applications and Examples

Special Case

The central limit theorem also works with weighted means (under the Lyapunov condition) even if heavy weights are put on the first few observations, and weights are slowly decaying to zero. Depending on how fast weights are decaying to zero, it may or may not yield convergence to the normal distribution. When it convergences, and when it does not, is discussed in this article where five cases are illustrated with pictures: three that lead to convergence, and two that do not.

Simple Example

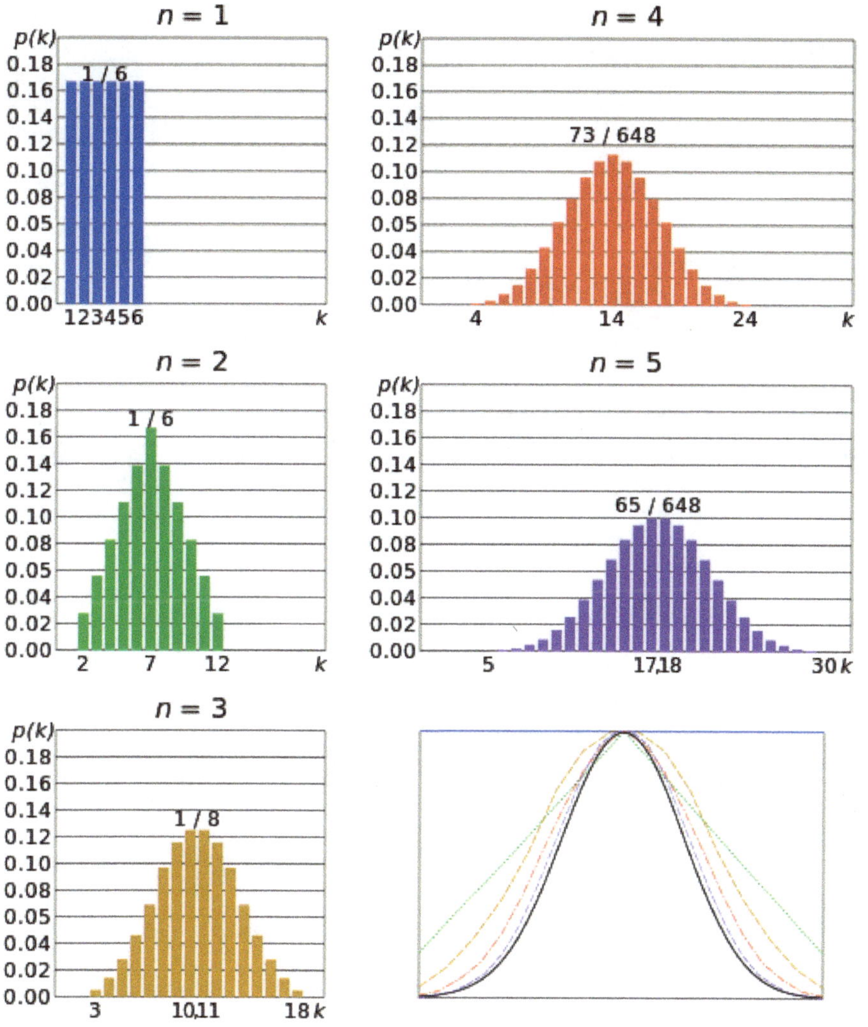

Comparison of probability density functions, **p(k) for the sum of n fair 6-sided dice to show their convergence to a normal distribution with increasing n, in accordance to the central limit theorem. In the bottom-right graph, smoothed profiles of the previous graphs are rescaled, superimposed and compared with a normal distribution (black curve).

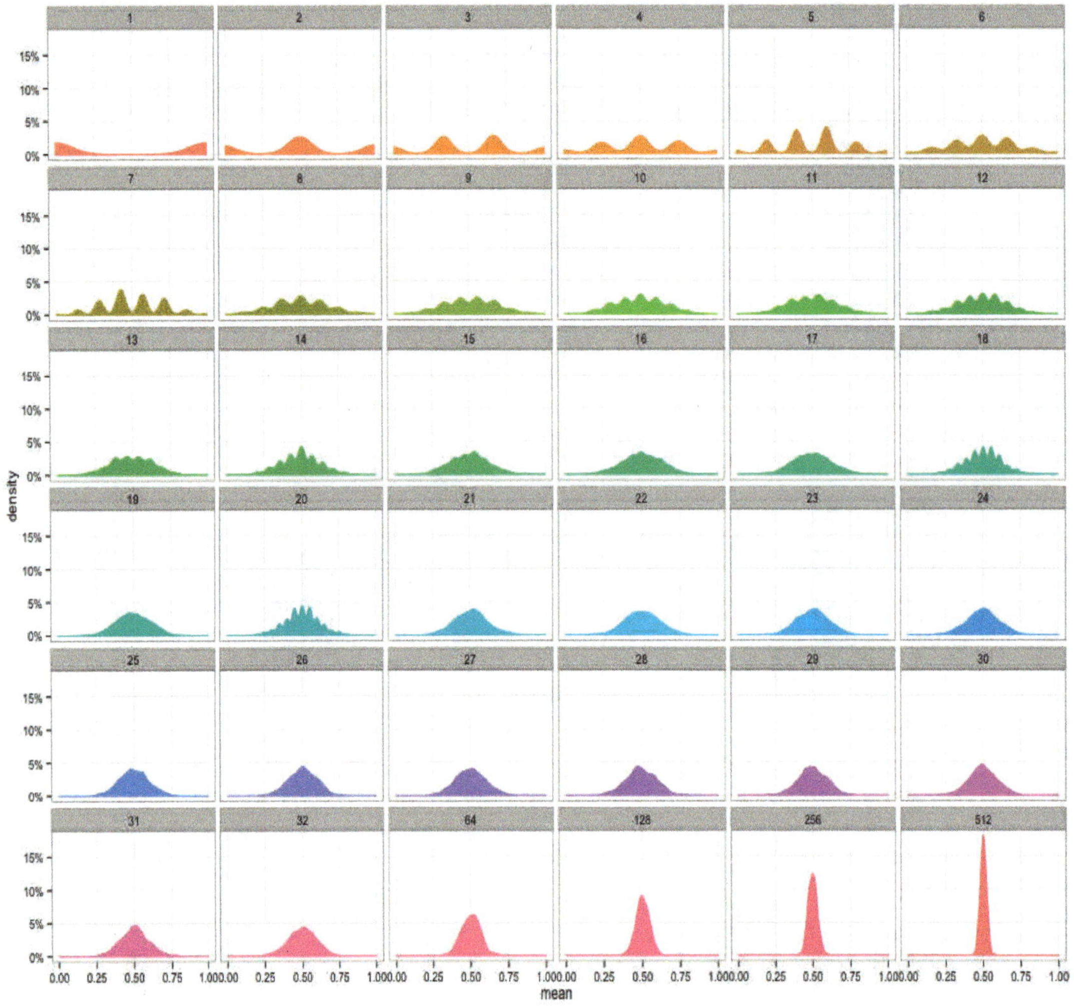

Another simulation using the binomial distribution. Random 0s and 1s were generated, and then their means calculated for sample sizes ranging from 1 to 512. Note that as the sample size increases the tails become thinner and the distribution becomes more concentrated around the mean.

A simple example of the central limit theorem is rolling a large number of identical, unbiased dice. The distribution of the sum (or average) of the rolled numbers will be well approximated by a normal distribution. Since real-world quantities are often the balanced sum of many unobserved random events, the central limit theorem also provides a partial explanation for the prevalence of the normal probability distribution. It also justifies the approximation of large-sample statistics to the normal distribution in controlled experiments.

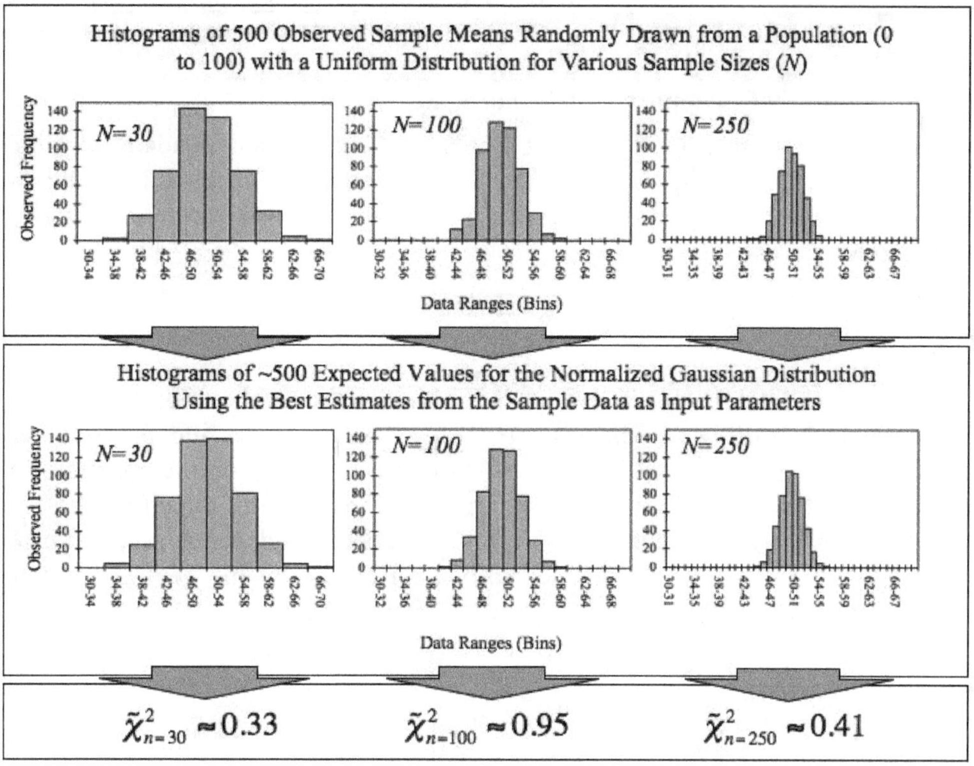

This figure demonstrates the central limit theorem. The sample means are generated using a random number generator, which draws numbers between 0 and 100 from a uniform probability distribution. It illustrates that increasing sample sizes result in the 500 measured sample means being more closely distributed about the population mean (50 in this case). It also compares the observed distributions with the distributions that would be expected for a normalized Gaussian distribution, and shows the chi-squared values that quantify the goodness of the fit (the fit is good if the reduced chi-squared value is less than or approximately equal to one). The input into the normalized Gaussian function is the mean of sample means (~50) and the mean sample standard deviation divided by the square root of the sample size (~28.87/√n), which is called the standard deviation of the mean (since it refers to the spread of sample means).

Real Applications

Published literature contains a number of useful and interesting examples and applications relating to the central limit theorem. One source states the following examples:

- The probability distribution for total distance covered in a random walk (biased or unbiased) will tend toward a normal distribution.
- Flipping a large number of coins will result in a normal distribution for the total number of heads (or equivalently total number of tails).

From another viewpoint, the central limit theorem explains the common appearance of the Bell Curve" in density estimates applied to real world data. In cases like electronic noise, examination grades, and so on, we can often regard a single measured value as the weighted average of a large number of small effects. Using generalisations of the central limit theorem, we can then see that this would often (though not always) produce a final distribution that is approximately normal.

In general, the more a measurement is like the sum of independent variables with equal influence on the result, the more normality it exhibits. This justifies the common use of this distribution to stand in for the effects of unobserved variables in models like the linear model.

Regression

Regression analysis and in particular ordinary least squares specifies that a dependent variable depends according to some function upon one or more independent variables, with an additive error term. Various types of statistical inference on the regression assume that the error term is normally distributed. This assumption can be justified by assuming that the error term is actually the sum of a large number of independent error terms; even if the individual error terms are not normally distributed, by the central limit theorem their sum can be well approximated by a normal distribution.

Other Illustrations

Given its importance to statistics, a number of papers and computer packages are available that demonstrate the convergence involved in the central limit theorem.

Dutch mathematician Henk Tijms writes:

The central limit theorem has an interesting history. The first version of this theorem was postulated by the French-born mathematician Abraham de Moivre who, in a remarkable article published in 1733, used the normal distribution to approximate the distribution of the number of heads resulting from many tosses of a fair coin. This finding was far ahead of its time, and was nearly forgotten until the famous French mathematician Pierre-Simon Laplace rescued it from obscurity in his monumental work Théorie analytique des probabilités, which was published in 1812. Laplace expanded De Moivre's finding by approximating the binomial distribution with the normal distribution. But as with De Moivre, Laplace's finding received little attention in his own time. It was not until the nineteenth century was at an end that the importance of the central limit theorem was discerned, when, in 1901, Russian mathematician Aleksandr Lyapunov defined it in general terms and proved precisely how it worked mathematically. Nowadays, the central limit theorem is considered to be the unofficial sovereign of probability theory.

Sir Francis Galton described the Central Limit Theorem in this way:

I know of scarcely anything so apt to impress the imagination as the wonderful form of cosmic order expressed by the Law of Frequency of Error". The law would have been personified by the Greeks and deified, if they had known of it. It reigns with serenity and in complete self-effacement, amidst the wildest confusion. The huger the mob, and the greater the apparent anarchy, the more perfect is its sway. It is the supreme law of Unreason. Whenever a large sample of chaotic elements are taken in hand and marshalled in the order of their magnitude, an unsuspected and most beautiful form of regularity proves to have been latent all along.

The actual term central limit theorem" (in German: zentraler Grenzwertsatz") was first used by George Pólya in 1920 in the title of a paper. Pólya referred to the theorem as central" due to its importance in probability theory. According to Le Çam, the French school of probability interprets the word central in the sense that "it describes the behaviour of the centre of the distribution as opposed to its tails".The abstract of the paper On the central limit theorem of calculus of probability and the problem of moments by Pólya in 1920 translates as follows.

The occurrence of the Gaussian probability density $1 = e^{-x2}$ in repeated experiments, in errors of measurements, which result in the combination of very many and very small elementary errors, in diffusion processes etc., can be explained, as is well-known, by the very same limit theorem, which plays a central role in the calculus of probability. The actual discoverer of this limit theorem is to be named Laplace; it is likely that its rigorous proof was first given by Tschebyscheff and its sharpest formulation can be found, as far as I am aware of, in an article by Liapounoff. [...]

A thorough account of the theorem`s history, detailing Laplace`s foundational work, as well as Cauchy`s, Bessel`s and Poisson`s contributions, is provided by Hald. Two historical accounts, one covering the development from Laplace to Cauchy, the second the contributions by von Mises, Pólya, Lindeberg, Lévy, and Cramér during the 1920s, are given by Hans Fischer. Le Cam describes a period around 1935. Bernstein presents a historical discussion focusing on the work of Pafnuty Chebyshev and his students Andrey Markov and Aleksandr Lyapunov that led to the first proofs of the CLT in a general setting.

A curious footnote to the history of the Central Limit Theorem is that a proof of a result similar to the 1922 Lindeberg CLT was the subject of Alan Turing`s 1934 Fellowship Dissertation for King`s College at the University of Cambridge. Only after submitting the work did Turing learn it had already been proved. Consequently, Turing`s dissertation was never published.

REFERENCES

1. http://www.math.uah.edu/stat/sample/CLT.html
2. Rice, John (1995), Mathematical Statistics and Data Analysis (Second ed.), Duxbury Press, ISBN 0-534-20934-3).
3. Voit, Johannes (2003), The Statistical Mechanics of Financial Markets, Springer-Verlag, p. 124, ISBN 3-540-00978-7
4. Billingsley (1995, p. 357)
5. Bauer (2001, Theorem 30.13, p.199)
6. Billingsley (1995, p.362)
7. Van der Vaart, A. W. (1998), Asymptotic statistics, New York: Cambridge University Press, ISBN 978-0-521-49603-2, LCCN 98015176
8. Voit, Johannes (2003). Section 5.4.3". The Statistical Mechanics of Financial Markets. Texts and Monographs in Physics. Springer-Verlag. ISBN 3-540-00978-7.

9. Gnedenko, B. V.; Kolmogorov, A. N. (1954). Limit distributions for sums of independent random variables. Cambridge: Addison-Wesley.

10. Uchaikin, Vladimir V.; Zolotarev, V. M. (1999). Chance and stability: stable distributions and their applications. VSP. pp. 61–62. ISBN 90-6764-301-7.

11. Billingsley (1995, Theorem 27.4)

12. Durrett (2004, Sect. 7.7(c), Theorem 7.8)

13. Durrett (2004, Sect. 7.7, Theorem 7.4)

14. Billingsley (1995, Theorem 35.12)

15. An Introduction to Stochastic Processes in Physics". jhupbooks.press.jhu.edu. Retrieved 2016-08-11.

16. Stein, C. (1972), A bound for the error in the normal approximation to the distribution of a sum of dependent random variables", Proceedings of the Sixth Berkeley Symposium on Mathematical Statistics and Probability: 583–602, MR 402873, Zbl 0278.60026

17. Chen, L.H.Y.; Goldstein, L.; and Shao, Q.M (2011), Normal approximation by Stein`s method, Springer, ISBN 978-3-642-15006-7

18. Artstein, S.; Ball, K.; Barthe, F.; Naor, A. (2004), Solution of Shannon`s Problem on the Monotonicity of Entropy", Journal of the American Mathematical Society, ***17*** (4): 975–982, doi:10.1090/S0894-0347-04-00459-X

19. Rosenthal, Jeffrey Seth (2000) A first look at rigorous probability theory, World Scientific, ISBN 981-02-4322-7.(Theorem 5.3.4, p. 47)

20. Johnson, Oliver Thomas (2004) Information theory and the central limit theorem, Imperial College Press, 2004, ISBN 1-86094-473-6. (p. 88)

21. Borodin, A. N. ; Ibragimov, Il`dar Abdulovich; Sudakov, V. N. (1995) Limit theorems for functionals of random walks, AMS Bookstore, ISBN 0-8218-0438-3. (Theorem 1.1, p. 8)

22. Petrov, V.V. (1976), 7", Sums of Independent Random Variables, New York-Heidelberg: Springer-Verlag

23. Rempala, G.; Wesolowski, J. (2002). Asymptotics of products of sums and U-statistics" (PDF). Electronic Communications in Probability. *7*: 47–54. doi:10.1214/ecp.v7-1046.

24. Klartag (2007, Theorem 1.2)

25. Durrett (2004, Section 2.4, Example 4.5)

26. Klartag (2008, Theorem 1)

27. Klartag (2007, Theorem 1.1)

28. Zygmund, Antoni (1959), Trigonometric series, Volume II, Cambridge. (2003 combined volume I,II: ISBN 0-521-89053-5) (Sect. XVI.5, Theorem 5-5)

29. Gaposhkin (1966, Theorem 2.1.13)

30. Bárány & Vu (2007, Theorem 1.1)

31. Bárány & Vu (2007, Theorem 1.2)

32. Meckes, Elizabeth (2008), Linear functions on the classical matrix groups", Transactions of the American Mathematical Society, ***360*** (10): 5355–5366, arXiv:math/0509441, doi:10.1090/S0002-9947-08-04444-9

33. Gaposhkin (1966, Sect. 1.5)

34. Kotani, M.; Sunada, Toshikazu (2003). Spectral geometry of crystal lattices. ***338***. Contemporary Math. pp. 271–305. ISBN 978-0-8218-4269-0.

35. Sunada, Toshikazu (2012). Topological Crystallography ---With a View Towards Discrete Geometric Analysis---. Surveys and Tutorials in the Applied Mathematical Sciences. ***6***. Springer. ISBN 978-4-431-54177-6.

36. Dinov, Christou & Sanchez (2008)

37. SOCR CLT Activity wiki

38. Marasinghe, M., Meeker, W., Cook, D. & Shin, T.S.(1994 August), Using graphics and simulation to teach statistical concepts", Paper presented at the Annual meeting of the American Statistician Association, Toronto, Canada.

39. Henk, Tijms (2004), Understanding Probability: Chance Rules in Everyday Life, Cambridge: Cambridge University Press, p. 169, ISBN 0-521-54036-4

40. Galton F. (1889) Natural Inheritance , p. 66

41. Pólya, George (1920), Über den zentralen Grenzwertsatz der Wahrscheinlichkeitsrechnung und das Momentenproblem", Mathematische Zeitschrift (in German), ***8*** (3–4): 171–181, doi:10.1007/BF01206525

42. Le Cam, Lucien (1986), The central limit theorem around 1935", Statistical Science, ***1*** (1): 78–91, doi:10.2307/2245503

43. Hald, Andreas A History of Mathematical Statistics from 1750 to 1930, Ch.17.

44. Fischer, Hans (2011), A History of the Central Limit Theorem: From Classical to Modern Probability Theory, Sources and Studies in the History of Mathematics and Physical Sciences, New York: Springer, doi:10.1007/978-0-387-87857-7, ISBN 978-0-387-87856-0, MR 2743162, Zbl 1226.60004 (Chapter 2: The Central Limit Theorem from Laplace to Cauchy: Changes in Stochastic Objectives and in Analytical Methods, Chapter 5.2: The Central Limit Theorem in the Twenties)

45. Bernstein, S.N. (1945) On the work of P.L.Chebyshev in Probability Theory, Nauchnoe Nasledie P.L.Chebysheva. Vypusk Pervyi: Matematika. (Russian) [The Scientific Legacy of P. L. Chebyshev. First Part: Mathematics, Edited by S. N. Bernstein.] Academiya Nauk SSSR, Moscow-Leningrad, 174 pp.

46. Hodges, Andrew (1983) Alan Turing: the enigma. London: Burnett Books., pp. 87-88.

47. Zabell, S.L. (2005) Symmetry and its discontents: essays on the history of inductive probability, Cambridge University Press. ISBN 0-521-44470-5. (pp. 199 ff.)

48. Aldrich, John (2009) "England and Continental Probability in the Inter-War Years", Electronic Journ@l for History of Probability and Statistics, vol. 5/2, Decembre 2009. (Section 3)

49. Jørgensen, Bent (1997). The theory of dispersion models. Chapman & Hall. ISBN 978-0412997112.

CHAPTER 8

Venn Diagrams

Chapter Contents

- **Introduction**
- **History**
- **Overview**
- **Randolph Diagram**
- **Logical Connective**
- **In Language**
- **Octahedron**

INTRODUCTION

A **Venn diagram** is a diagram that shows all possible logical relations between a finite collection of different sets. These diagrams depict elements as points in the plane, and sets as regions inside closed curves. A Venn diagram consists of multiple overlapping closed curves, usually circles, each representing a set. The points inside a curve labelled S represent elements of the set S, while points outside the boundary represent elements not in the set S. Thus, for example, the set of all elements that are members of both sets S and T, $S \cap T$, is represented visually by the area of overlap of the regions S and T. In Venn diagrams the curves are overlapped in every possible way, showing all possible relations between the sets. They are thus a special case of Euler diagrams, which do not necessarily show all relations. Venn diagrams were conceived around 1880 by John Venn. They are used to teach elementary set theory, as well as illustrate simple set relationships in probability, logic, statistics, linguistics and computer science.

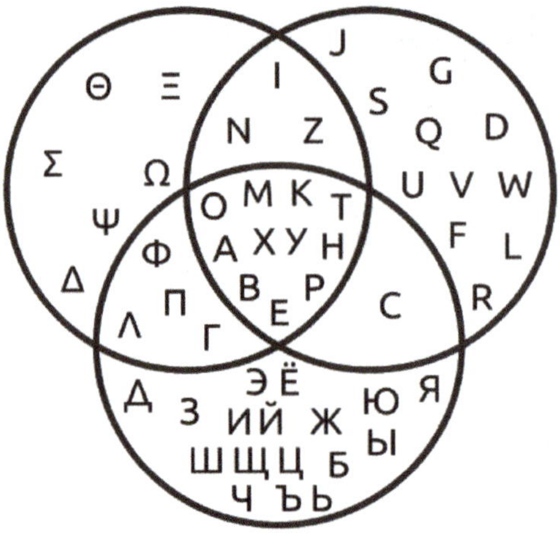

Venn diagram showing which uppercase letter glyphs are shared by the Greek, Latin and Russian alphabets

A Venn diagram in which in addition the area of each shape is proportional to the number of elements it contains is called an **area-proportional** or **scaled Venn diagram**.

Example

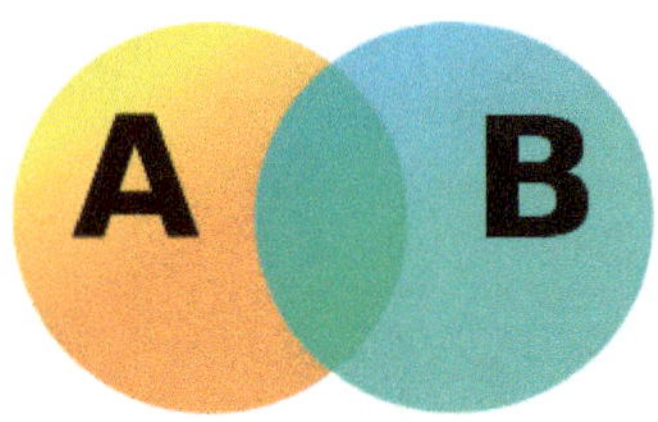

Sets A (creatures with two legs) and B (creatures that can fly).

This example involves two sets, A and B, represented here as coloured circles. The orange circle, set A, represents all living creatures that are two-legged. The blue circle, set B, represents the living creatures that can fly. Each separate type of creature can be imagined as a point somewhere in the diagram. Living creatures that both can fly *and* have two legs—for example, parrots—are then in both sets, so they correspond to points in the region where the blue and orange circles overlap. That region contains all such and only such living creatures.

Humans and penguins are bipedal, and so are then in the orange circle, but since they cannot fly they appear in the left part of the orange circle, where it does not overlap with the blue circle. Mosquitoes have six legs, and fly, so the point for mosquitoes is in the part of the blue circle that does not overlap with the orange one. Creatures that are not two-legged and cannot fly (for example, whales and spiders) would all be represented by points outside both circles.

The combined region of sets A and B is called the *union* of A and B, denoted by A ∪ B. The union in this case contains all living creatures that are either two-legged or that can fly (or both).

The region in both A and B, where the two sets overlap, is called the *intersection* of A and B, denoted by A ∩ B. For example, the intersection of the two sets is not empty, because there *are* points that represent creatures that are in *both* the orange and blue circles.

HISTORY

Venn diagrams were introduced in 1880 by John Venn in a paper entitled *On the Diagrammatic and Mechanical Representation of Propositions and Reasonings* in the "Philosophical Magazine and Journal of Science", about the different ways to represent propositions by diagrams. The use of these types of diagrams in formal logic, according to Ruskey and M. Weston, is "not an easy history to trace, but it is certain that the diagrams that are popularly associated with Venn, in fact, originated much earlier. They are rightly associated with Venn, however, because he comprehensively surveyed and formalized their usage, and was the first to generalize them".

Venn himself did not use the term "Venn diagram" and referred to his invention as "Eulerian Circles". For example, in the opening sentence of his 1880 article Venn writes, "Schemes of diagrammatic representation have been so familiarly introduced into logical treatises during the last century or so, that many readers, even those who have made no professional study of logic, may be supposed to be acquainted with the general nature and object of such devices. Of these schemes one only, viz. that commonly called ‹Eulerian circles,› has met with any general acceptance..." The first to use the term "Venn diagram" was Clarence Irving Lewis in 1918, in his book "A Survey of Symbolic Logic".

Venn diagrams are very similar to Euler diagrams, which were invented by Leonhard Euler in the 18th century M. E. Baron has noted that Leibniz (1646–1716) in the 17th century produced similar diagrams before Euler, but much of it was unpublished. She also observes even earlier Euler-like diagrams by Ramon Lull in the 13th Century.

In the 20th century, Venn diagrams were further developed. D.W. Henderson showed in 1963 that the existence of an n-Venn diagram with n-fold rotational symmetry implied that n was a prime number. He also showed that such symmetric Venn diagrams exist when n is 5 or 7. In 2002 Peter Hamburger found symmetric Venn diagrams for $n = 11$ and in 2003, Griggs, Killian, and Savage showed that symmetric Venn diagrams exist for all other primes. Thus rotationally symmetric Venn diagrams exist if and only if n is a prime number.

Venn diagrams and Euler diagrams were incorporated as part of instruction in set theory as part of the new math movement in the 1960s. Since then, they have also been adopted in the curriculum of other fields such as reading.

OVERVIEW

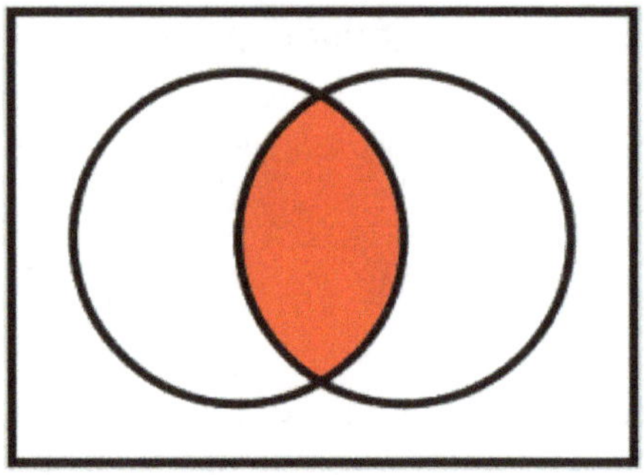

Intersection of two sets
$A \cap B$

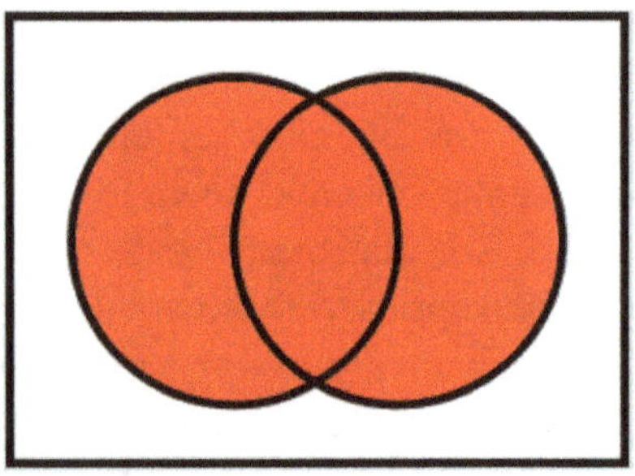

Union of two sets
$A \cup B$

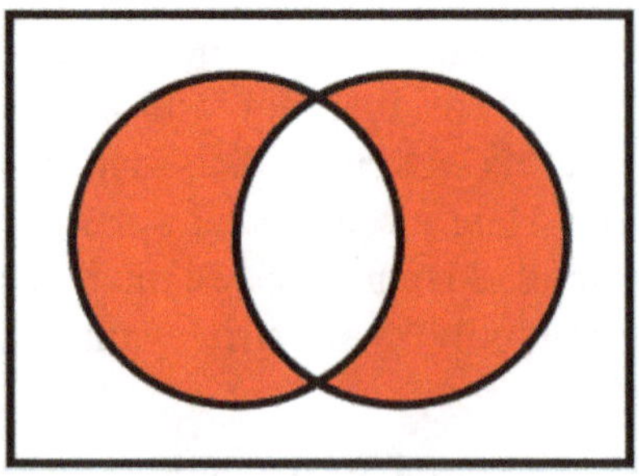

Symmetric difference of two sets $A \, \Delta \, B$

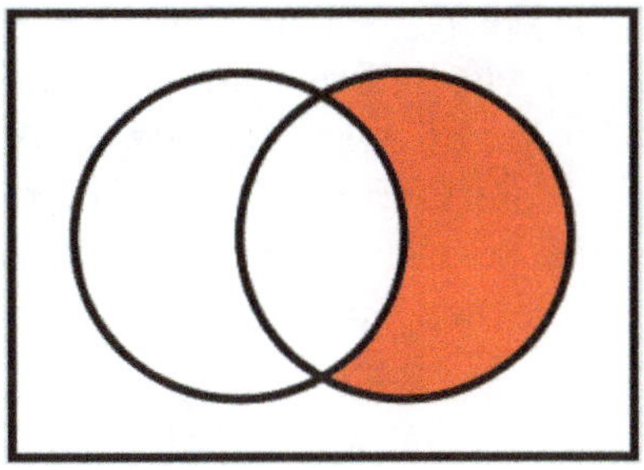

Relative complement of A (left) in B (right)
$A^c \cap B = B \setminus A$

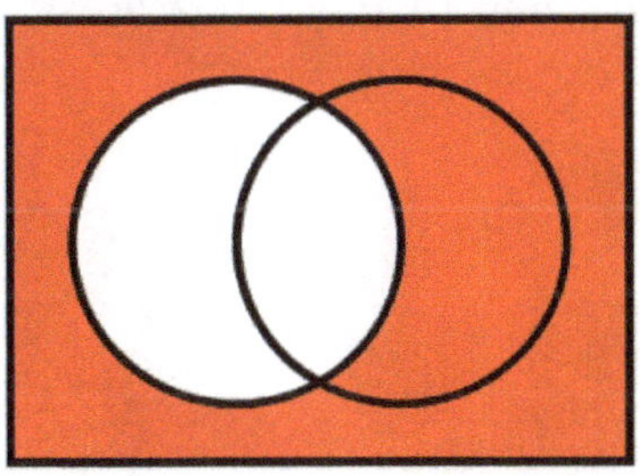

Absolute complement of A in U $A^c = U \setminus A$

A Venn diagram is constructed with a collection of simple closed curves drawn in a plane. According to Lewis, the "principle of these diagrams is that classes [or *sets*] be represented by regions in such relation to one another that all the possible logical relations of these classes can be indicated in the same diagram. That is, the diagram initially leaves room for any possible relation of the classes, and the actual or given relation, can then be specified by indicating that some particular region is null or is not-null".

Venn diagrams normally comprise overlapping circles. The interior of the circle symbolically represents the elements of the set, while the exterior represents elements that are not members of the set. For instance, in a two-set Venn diagram, one circle may represent the group of all wooden objects, while another circle may represent the set of all tables. The overlapping region or *intersection* would then represent the set of all wooden tables. Shapes other than circles can be employed as shown below by Venn's own higher set diagrams. Venn diagrams do not generally contain information on the relative or absolute sizes (cardinality) of sets; i.e. they are schematic diagrams.

Venn diagrams are similar to Euler diagrams. However, a Venn diagram for n component sets must contain all 2^n hypothetically possible zones that correspond to some combination of inclusion or exclusion in each of the component sets. Euler diagrams contain only the actually possible zones in a given context. In Venn diagrams, a shaded zone may represent an empty zone, whereas in an Euler diagram the corresponding zone is missing from the diagram. For example, if one set represents *dairy products* and another *cheeses*, the Venn diagram contains a zone for cheeses that are not dairy products. Assuming that in the context *cheese* means some type of dairy product, the Euler diagram has the cheese zone entirely contained within the dairy-product zone—there is no zone for (non-existent) non-dairy cheese. This means that as the number of contours increases, Euler diagrams are typically less visually complex than the equivalent Venn diagram, particularly if the number of non-empty intersections is small.

The difference between Euler and Venn diagrams can be seen in the following example. Take the three sets:

- $A = \{1, 2, 5\}$
- $B = \{1, 6\}$
- $C = \{4, 7\}$

The Venn and the Euler diagram of those sets are:

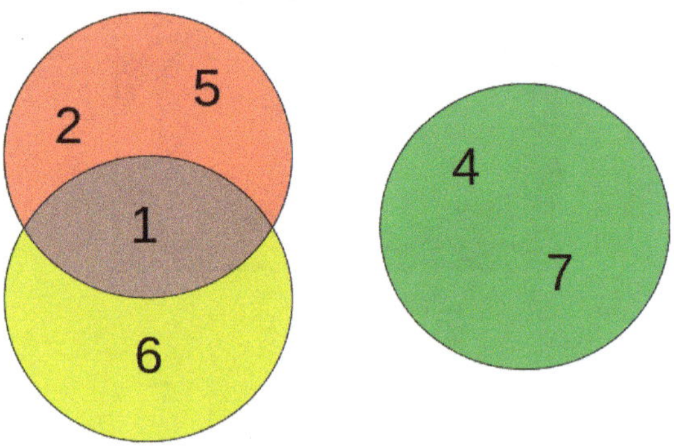

Euler diagram.

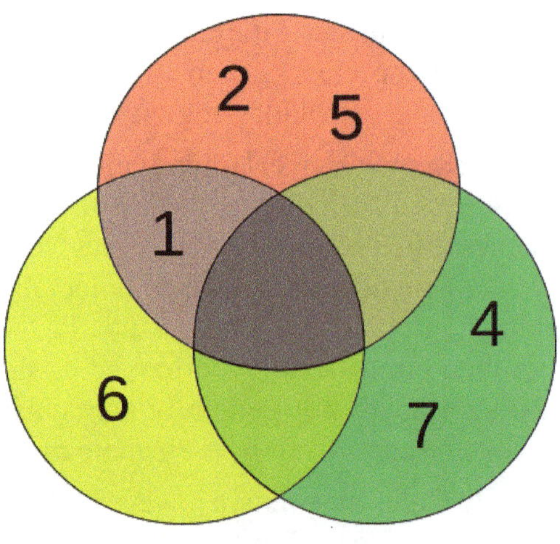

Venn diagram.

Extensions to Higher Numbers of Sets

Venn diagrams typically represent two or three sets, but there are forms that allow for higher numbers. Shown below, four intersecting spheres form the highest order Venn diagram that has the symmetry of a simplex and can be visually represented. The 16 intersections correspond to the vertices of a tesseract (or the cells of a 16-cell respectively).

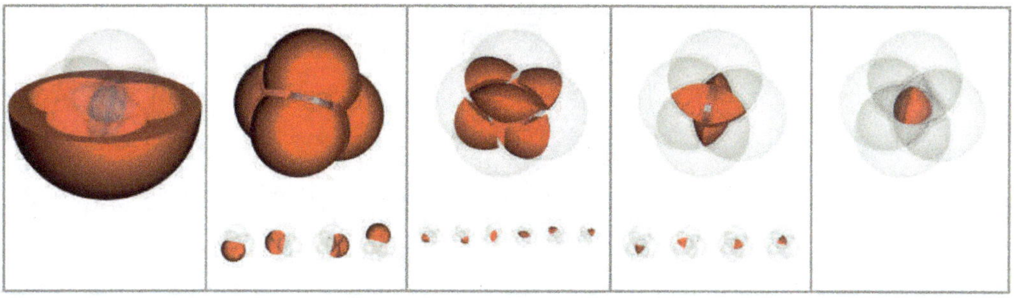

For higher numbers of sets, some loss of symmetry in the diagrams is unavoidable. Venn was keen to find "symmetrical figures...elegant in themselves," that represented higher numbers of sets, and he devised a four-set diagram using ellipses (see below). He also gave a construction for Venn diagrams for *any* number of sets, where each successive curve that delimits a set interleaves with previous curves, starting with the three-circle diagram.

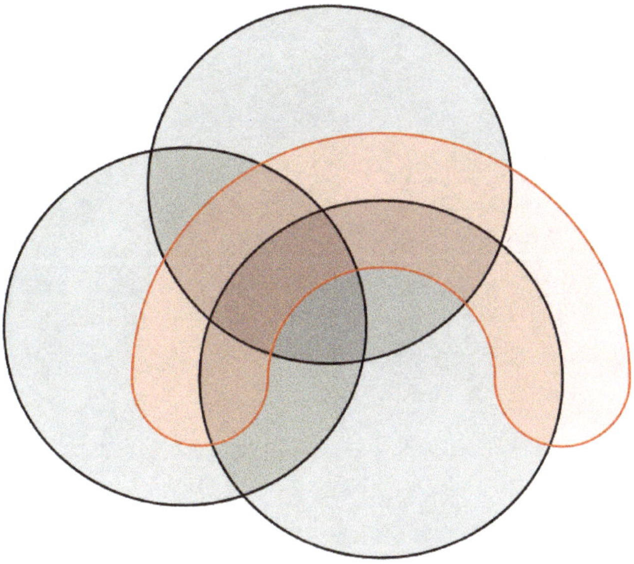

Venn's construction for 4 sets

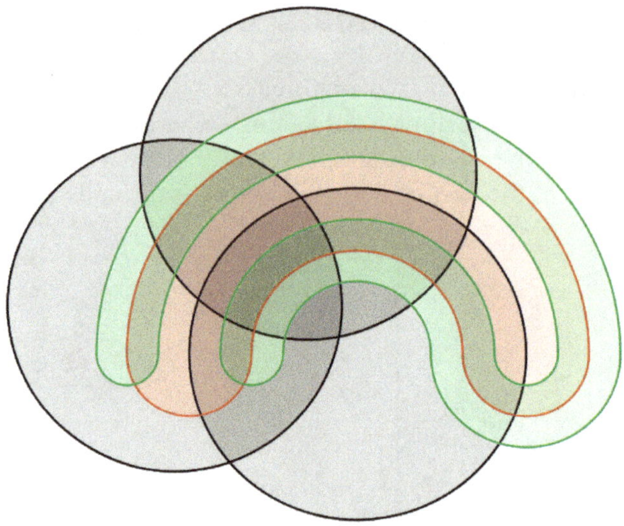

Venn's construction for 5 sets

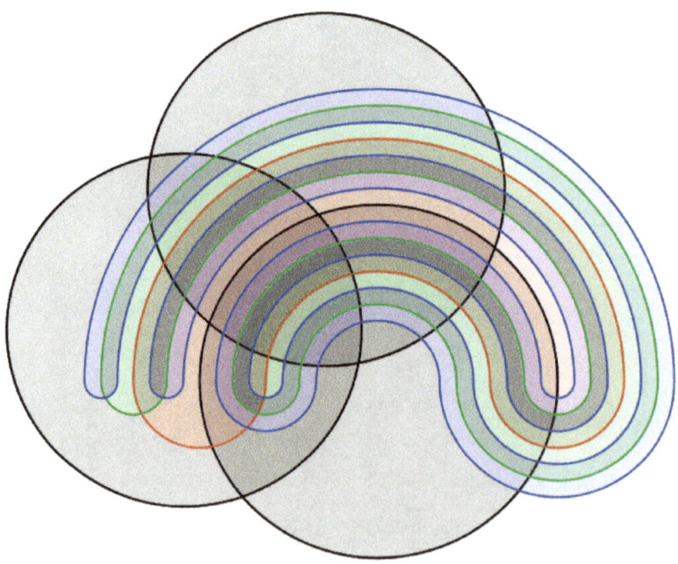

Venn's construction for 6 sets

Venn's four-set diagram using ellipses

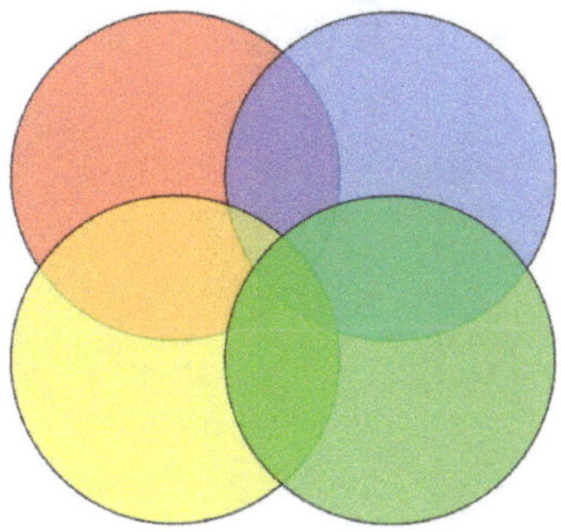

Non-example: This Euler diagram is **not** a Venn diagram for four sets as it has only 13 regions (excluding the outside); there is no region where only the yellow and blue, or only the red and green circles meet.

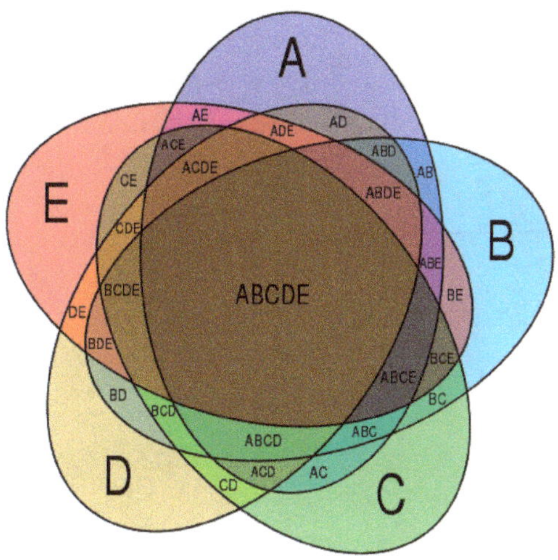

Five-set Venn diagram using congruent ellipses in a 5-fold rotationally symmetrical arrangement devised by Branko Grünbaum. Labels have been simplified for greater readability; for example, **A** denotes $\mathbf{A} \cap \mathbf{B}^c \cap \mathbf{C}^c \cap \mathbf{D}^c \cap \mathbf{E}^c$, while **BCE** denotes $\mathbf{A}^c \cap \mathbf{B} \cap \mathbf{C} \cap \mathbf{D}^c \cap \mathbf{E}$.

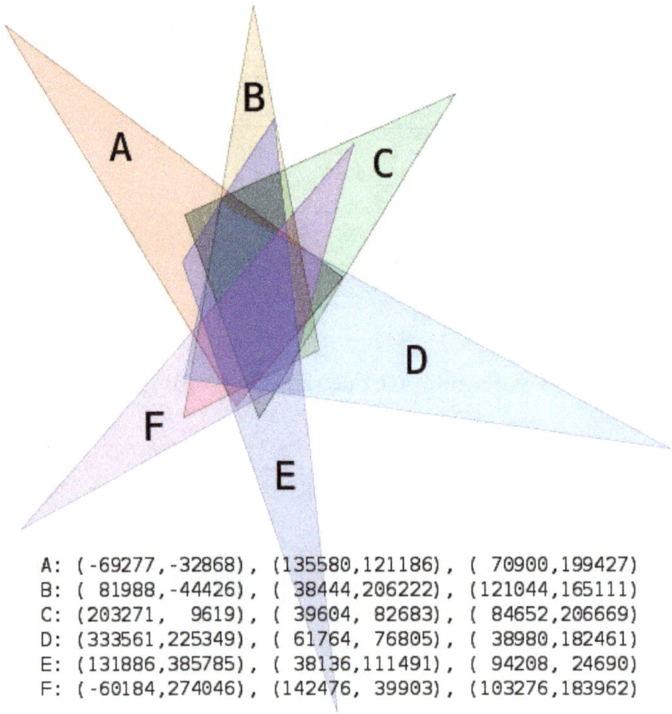

Six-set Venn diagram made of only triangles (interactive version)

Edwards' Venn diagrams

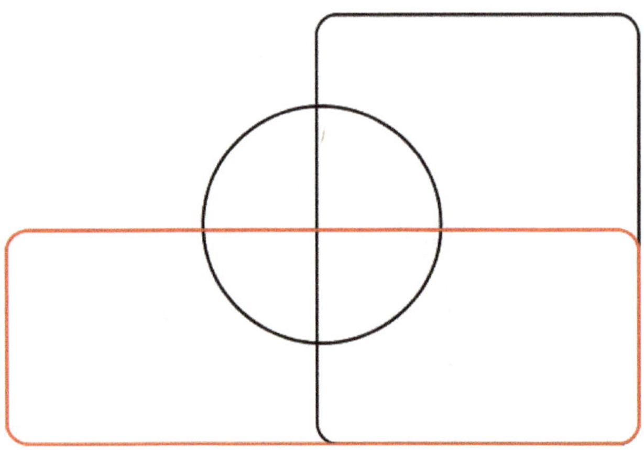

Three sets

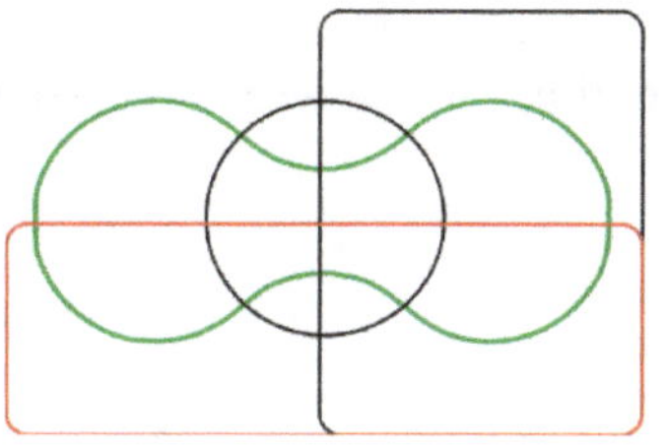

Four sets

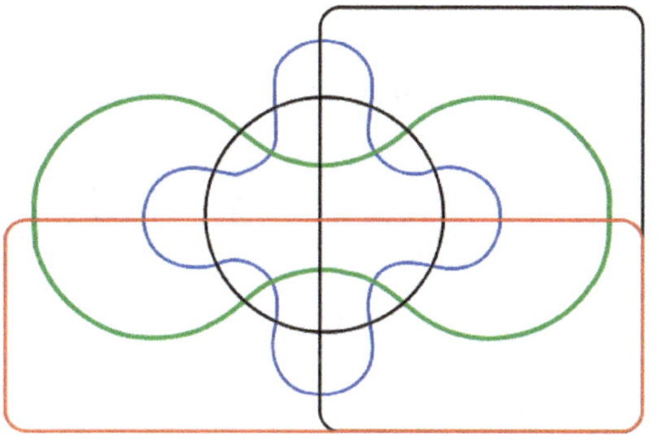

Five sets

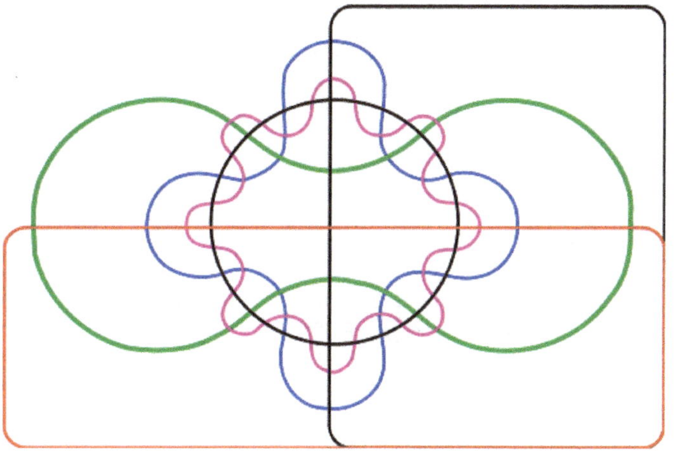

Six sets.

A. W. F. Edwards constructed a series of Venn diagrams for higher numbers of sets by segmenting the surface of a sphere. For example, three sets can be easily represented by taking three hemispheres of the sphere at right angles (x = 0, y = 0 and z = 0). A fourth set can be added to the representation by taking a curve similar to the seam on a tennis ball, which winds up and down around the equator, and so on. The resulting sets can then be projected back to a plane to give *cogwheel* diagrams with increasing numbers of teeth, as shown here. These diagrams were devised while designing a stained-glass window in memory of Venn.

Other Diagrams

Edwards' Venn diagrams are topologically equivalent to diagrams devised by Branko Grünbaum, which were based around intersecting polygons with increasing numbers of sides. They are also 2-dimensional representations of hypercubes.

Henry John Stephen Smith devised similar *n*-set diagrams using sine curves with the series of equations

$$y_i = \frac{\sin(2^i x)}{2i} \text{ where } 0 \leq i \leq n-2 \text{ and } i \in \mathbb{N}.$$

Charles Lutwidge Dodgson devised a five-set diagram.

Related Concepts

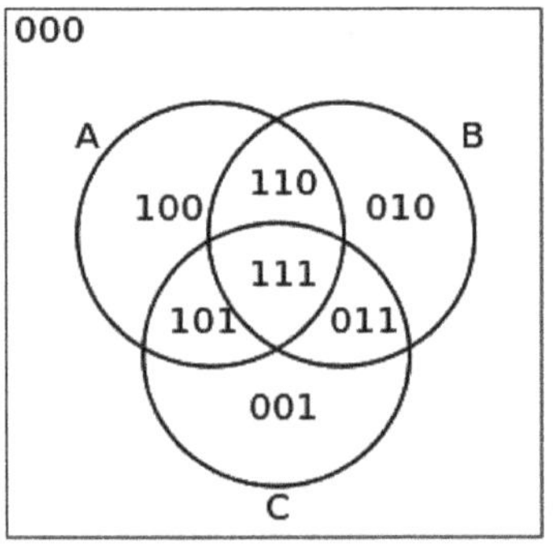

Venn diagram as a truth table

Venn diagrams correspond to truth tables for the propositions $x \in A, x \in B$, etc., in the sense that each region of Venn diagram corresponds to one row of the truth table. Another way of representing sets is with R-Diagrams.

RANDOLPH DIAGRAM

Randolph diagram that represents the logical statement $P \vee Q$ (disjunction).

A **Randolph diagram (R-diagram)** is a simple way to visualize logical expressions and combinations of sets. Randolph diagrams were created by mathematician John F. Randolph in 1965, while he was teaching at the University of Arkansas.

Randolph diagrams can be interpreted most easily by defining each line as *belonging to* or *relating to* one logical statement or set. Any dot above the line indicates truth or inclusion and below the line indicates falsity or exclusion. Using this system, one can represent any combination of sets or logical statements using intersecting lines.

Though Venn diagrams are more commonly used to represent combinations of sets, Randolph diagrams have the advantage of being able to cleanly represent combinations of more than 3 sets. Venn diagrams require either extension into higher spatial dimensions or the use of more complicated shapes while Randolph diagrams evenly subdivide for every additional set. Here is a comparison between a Venn diagram and R-diagram for 5 sets or logical statements:

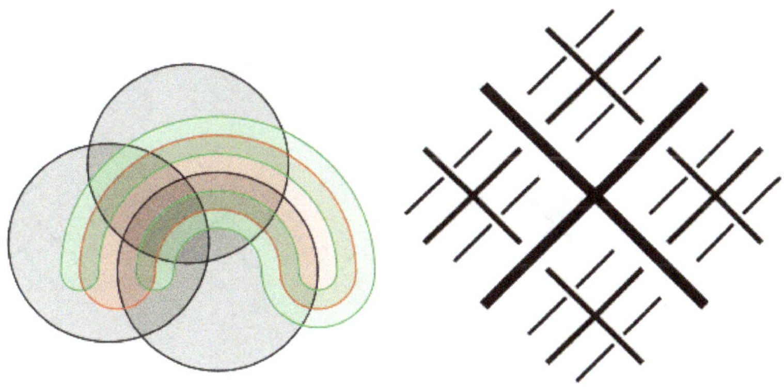

In his introductory paper on the subject, *Cross-Examining Propositional Calculus and Set Operations*, Randolph mentions that the first use of crosses and dots to represent logical relationships was introduced by W. S. McCulloch, a neurophysiologist and Randolph's contemporary. Randolph modified McCulloch's system with a new way of representing combinations and relationships of more than two logical statements or sets, namely subdividing each section of the R-diagram with a new diagonal line for each new element introduced. Randolph's paper suggests that his original notion was to use R-diagrams to represent logical relationships, and then expanded the idea to be applied to set theory as well. Throughout the paper, R-diagrams are used in conjunction with normal logical and set binary operation symbols.

Application to Logic Theory

When applying R-diagrams to logic theory, logical statements p, q, and r can each become a line or multiple lines to visually display the validity of each element in a larger statement. Generally, p is thought to be represented by an upward sloping line (/) while q is represented by a downward sloping line (\). A dot in the diagram above a slanting line indicates truth for that statement; likewise, a dot below indicates falsity. The R-diagrams for p and q are shown below, respectively:

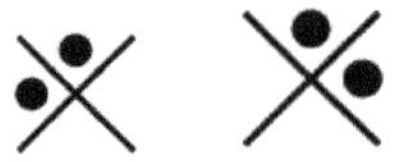

For more than two statements, the four spaces formed by the intersection of lines p and q must be subdivided into more lines. In the case of r, a single upward sloping line (/) is added in each of the four spaces. The R-diagram for r is shown below:

This method can be extended for any number of truth values:

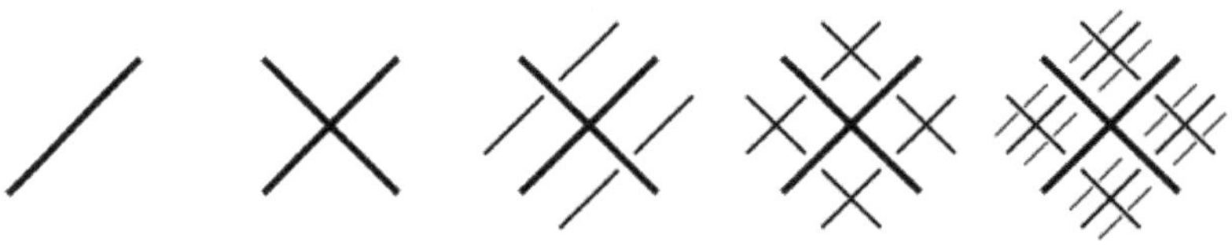

, etc.

R-diagrams are primarily used to represent logical expressions. Given a logical proposition, R-diagrams are able to display the outcome of every possible true/false variation of each element, creating an alternative way to represent a truth table.

Truth Table			
#	p	q	r
1	T	T	T
2	T	T	F
3	T	F	T
4	T	F	F
5	F	T	T
6	F	T	F
7	F	F	T
8	F	F	F

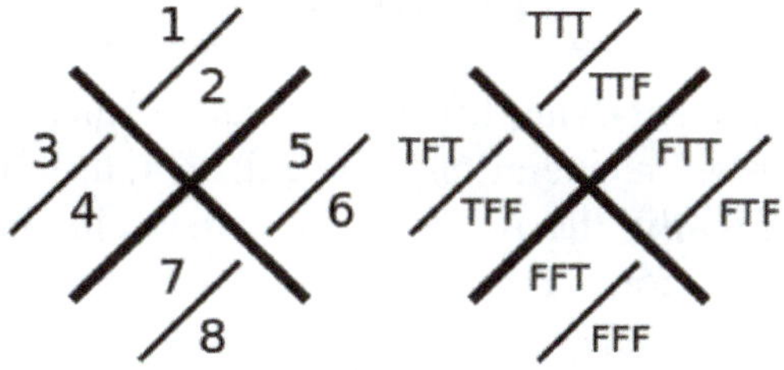

All the basic logical operations, or connectives, can be expressed using an R-diagrams as a more easily readable alternative to a truth table, as is shown in the table below:

Basic Logical Operations

<table>
<tr><th>Name</th><th>Symbols</th><th>R-Diagram</th><th>Truth Table</th></tr>
<tr><td>Negation (not)</td><td>¬, ~</td><td></td><td><table><tr><th>p</th><th>¬p</th></tr><tr><td>T</td><td>F</td></tr><tr><td>F</td><td>T</td></tr></table></td></tr>
<tr><td>Conjunction (and)</td><td>&, ∧</td><td></td><td><table><tr><th>p</th><th>q</th><th>p ∧ q</th></tr><tr><td>T</td><td>T</td><td>T</td></tr><tr><td>T</td><td>F</td><td>F</td></tr><tr><td>F</td><td>T</td><td>F</td></tr><tr><td>F</td><td>F</td><td>F</td></tr></table></td></tr>
<tr><td>Disjunction (or)</td><td>∨</td><td></td><td><table><tr><th>p</th><th>q</th><th>p ∨ q</th></tr><tr><td>T</td><td>T</td><td>T</td></tr><tr><td>T</td><td>F</td><td>T</td></tr><tr><td>F</td><td>T</td><td>T</td></tr><tr><td>F</td><td>F</td><td>F</td></tr></table></td></tr>
<tr><td>Material implication (if...then)</td><td>→, ⇒, ⊃</td><td></td><td><table><tr><th>p</th><th>q</th><th>p → q</th></tr><tr><td>T</td><td>T</td><td>T</td></tr><tr><td>T</td><td>F</td><td>F</td></tr><tr><td>F</td><td>T</td><td>T</td></tr><tr><td>F</td><td>F</td><td>T</td></tr></table></td></tr>
<tr><td>Biconditional (if and only if, xnor)</td><td>↔, ≡, =</td><td></td><td><table><tr><th>p</th><th>q</th><th>p ↔ q</th></tr><tr><td>T</td><td>T</td><td>T</td></tr><tr><td>T</td><td>F</td><td>F</td></tr><tr><td>F</td><td>T</td><td>F</td></tr><tr><td>F</td><td>F</td><td>T</td></tr></table></td></tr>
</table>

Simplifying Logical Expressions

R-diagrams can be used to easily simplify complicated logical expressions, using a step-by-step process. Using order of operations, logical operators are applied to R-diagrams in the proper sequence. Finally, the result is an R-diagram that can be converted back into a simpler logical expression.

For example, take the following expression:

$$(Q \leftrightarrow P) \vee (\neg P \wedge Q)$$

It can be simplified using R-diagrams as follows:

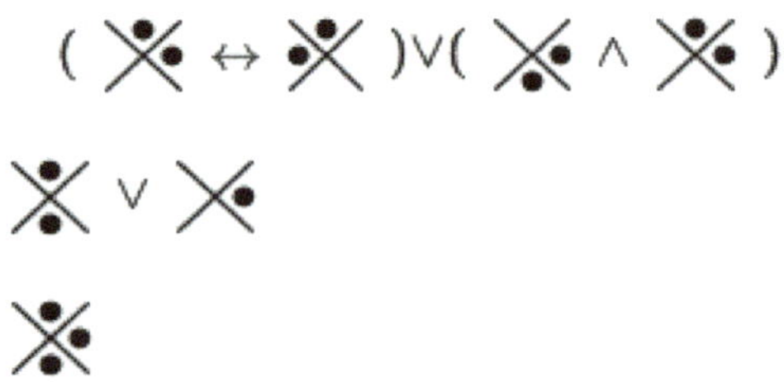

which is equal to:

$$P \to Q.$$

Proving Logical Arguments

Similarly, R-diagrams can be used to prove or disprove logical arguments. Take, for instance, the well known argument modus ponens, also known as implication elimination:

$$\frac{P \to Q, P}{\therefore Q}$$

This can be converted into a tautological logical expression,

$$((P \to Q) \wedge P) \to Q$$

which can then be simplified using R-diagrams:

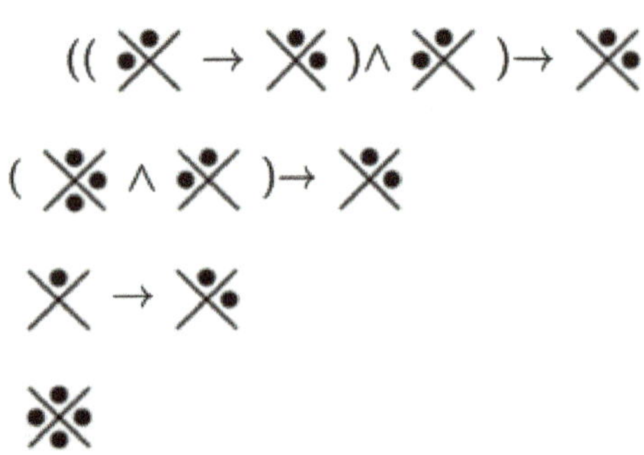

The result is an R-diagram in which every space has a dot. This means the argument is a tautology; it is true in all cases. An R-diagram in which no space has a dot is a contradiction, a statement that is never true.

Application to Set Theory

R-diagrams are also used in set theory, as an alternative to Venn diagrams. In set theory, each line represents a set instead of a logical statement; A replaces p and B replaces q. When used for sets, a dot above the line represents inclusion, where a dot below represents exclusion. As in logic, basic set operations can be represented visually using R-diagrams:

Name	Notation	R-Diagram
Union	$A \cup B$	
Intersection	$A \cap B$	
Absolute Complement	A^c	
Relative Complement (set difference)	$A \setminus B$	
Symmetric Difference	$A \Delta B$	

R-diagrams illustrate the equivalence between the set theoretical and logical concepts: intersection in set theory is equivalent to conjunction in logic, and set theory's union is equivalent to the logical disjunction.

LOGICAL CONNECTIVE

In logic, a **logical connective** (also called a **logical operator**) is a symbol or word used to connect two or more sentences (of either a formal or a natural language) in a grammatically valid way, such that the sense of the compound sentence produced depends only on the original sentences.

The most common logical connectives are **binary connectives** (also called **dyadic connectives**) which join two sentences which can be thought of as the function's operands. Also commonly, negation is considered to be a **unary connective**.

Logical connectives along with quantifiers are the two main types of logical constants used in formal systems such as propositional logic and predicate logic. Semantics of a logical connective is often, but not always, presented as a truth function.

A logical connective is similar to but not equivalent to a conditional operator.

IN LANGUAGE

Natural Language

In the grammar of natural languages two sentences may be joined by a grammatical conjunction to form a *grammatically* compound sentence. Some but not all such grammatical conjunctions are truth functions. For example, consider the following sentences:

A: Jack went up the hill.

B: Jill went up the hill.

C: Jack went up the hill *and* Jill went up the hill.

D: Jack went up the hill *so* Jill went up the hill.

The words *and* and *so* are *grammatical* conjunctions joining the sentences (A) and (B) to form the compound sentences (C) and (D). The *and* in (C) is a *logical* connective, since the truth of (C) is completely determined by (A) and (B): it would make no sense to affirm (A) and (B) but deny (C). However, *so* in (D) is not a logical connective, since it would be quite reasonable to affirm (A) and (B) but deny (D): perhaps, after all, Jill went up the hill to fetch a pail of water, not because Jack had gone up the hill at all.

Various English words and word pairs express logical connectives, and some of them are synonymous. Examples (with the name of the relationship in parentheses) are:

- "and" (conjunction)
- "and then" (conjunction with sequencing)
- "and then within" (conjunction with sequencing and time window requirement)
- "or" (disjunction)
- "either...or" (exclusive disjunction)
- "implies" (implication)
- "if...then" (implication)
- "if and only if" (equivalence)
- "only if" (implication)
- "just in case" (biconditional)
- "but" (conjunction)
- "however" (conjunction)
- "not both" (alternative denial)
- "neither...nor" (joint denial)

The word "not" (negation) and the phrases "it is false that" (negation) and "it is not the case that" (negation) also express a logical connective – even though they are applied to a single statement, and do not connect two statements.

Formal Languages

In formal languages, truth functions are represented by unambiguous symbols. These symbols are called "logical connectives", "logical operators", "propositional operators", or, in classical logic, "truth-functional connectives". See well-formed formula for the rules which allow new well-formed formulas to be constructed by joining other well-formed formulas using truth-functional connectives.

Logical connectives can be used to link more than two statements, so one can speak about "*n*-ary logical connective".

Common Logical Connectives

List of Common Logical Connectives

Name / Symbol		**Truth table**					**Venn diagram**
$P =$		0		1			
Truth/Tautology	⊤		1		1		
Proposition P			0		1		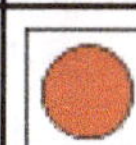
False/Contradiction	⊥		0		0		
Negation	¬		1		0		
Binary connectives		$P =$	0	0	1	1	
		$Q =$	0	1	0	1	
Conjunction	∧		0	0	0	1	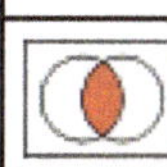
Alternative denial	↑		1	1	1	0	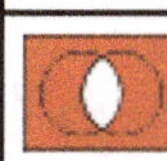
Disjunction	∨		0	1	1	1	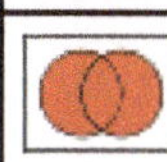
Joint denial	↓		1	0	0	0	

Material conditional	→		1	1	0	1	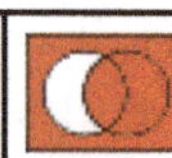
Exclusive or	↮		0	1	1	0	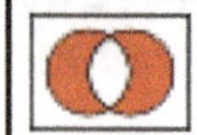
Biconditional	↔		1	0	0	1	
Converse implication	←		1	0	1	1	
Proposition P			0	0	1	1	
Proposition Q			0	1	0	1	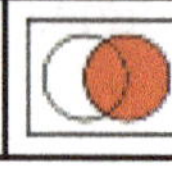
More information							

Commonly used logical connectives include

- Negation (not): ¬, N (prefix), ~
- Conjunction (and): ∧, K (prefix), &, ·
- Disjunction (or): ∨, A (prefix)
- Material implication (if...then): →, C (prefix), ⇒, ⊃
- Biconditional (if and only if): ↔, E (prefix), ≡, =

Alternative names for biconditional are "iff", "xnor" and "bi-implication".

For example, the meaning of the statements *it is raining* and *I am indoors* is transformed when the two are combined with logical connectives. For statement P = *It is raining* and Q = *I am indoors*:

- It is **not** raining ($\neg P$)
- It is raining **and** I am indoors ($\boldsymbol{P \wedge Q}$)
- It is raining **or** I am indoors ($\boldsymbol{P \vee Q}$)
- **If** it is raining, **then** I am indoors ($\boldsymbol{P \rightarrow Q}$)
- **If** I am indoors, **then** it is raining ($\boldsymbol{Q \rightarrow P}$)
- I am indoors **if and only if** it is raining ($\boldsymbol{P \leftrightarrow Q}$)

It is also common to consider the *always true* formula and the *always false* formula to be connective:

- True formula (⊤, 1, V [prefix], or T)
- False formula (⊥, 0, O [prefix], or F)

History of Notations

- Negation: the symbol ¬ appeared in Heyting in 1929. (compare to Frege›s symbol ⫟A in his Begriffsschrift); the symbol ~ appeared in Russell in 1908; an alternative notation is to add an horizontal line on top of the formula, as in $\overline{P}$ another alternative notation is to use a prime symbol as in P'.

- Conjunction: the symbol ∧ appeared in Heyting in 1929 (compare to Peano's use of the set-theoretic notation of intersection ∩); & appeared at least in Schönfinkel in 1924; **.** comes from Boole›s interpretation of logic as an elementary algebra.
- Disjunction: the symbol ∨ appeared in Russell in 1908 (compare to Peano›s use of the set-theoretic notation of union ∪); the symbol + is also used, in spite of the ambiguity coming from the fact that the + of ordinary elementary algebra is an exclusive or when interpreted logically in a two-element ring; punctually in the history a + together with a dot in the lower right corner has been used by Peirce,
- Implication: the symbol → can be seen in Hilbert in 1917; ⊃ was used by Russell in 1908 (compare to Peano's inverted C notation); ⇒ was used in Vax.
- Biconditional: the symbol ≡ was used at least by Russell in 1908; ↔ was used at least by Tarski in 1940; ⇔ was used in Vax; other symbols appeared punctually in the history such as ⊃⊂ in Gentzen, ~ in Schönfinkel or ⊂⊃ in Chazal.
- True: the symbol 1 comes from Boole's interpretation of logic as an elementary algebra over the two-element Boolean algebra; other notations include ⋀ to be found in Peano.
- False: the symbol 0 comes also from Boole›s interpretation of logic as a ring; other notations include ⋁ to be found in Peano.

Some authors used letters for connectives at some time of the history: **u.** for conjunction (German's "und" for "and") and **o.** for disjunction (German's "oder" for "or") in earlier works by Hilbert (1904); **N*p*** for negation, **K*pq*** for conjunction, **D*pq*** for alternative denial, **A*pq*** for disjunction, **X*pq*** for joint denial, **C*pq*** for implication, **E*pq*** for biconditional in Łukasiewicz (1929); cf. Polish notation.

Redundancy

Such a logical connective as converse implication "←" is actually the same as material conditional with swapped arguments; thus, the symbol for converse implication is redundant. In some logical calculi (notably, in classical logic) certain essentially different compound statements are logically equivalent. A less trivial example of a redundancy is the classical equivalence between $\neg P \vee Q$ and $P \rightarrow Q$. Therefore, a classical-based logical system does not need the conditional operator "→" if "¬" (not) and "∨" (or) are already in use, or may use the "→" only as a syntactic sugar for a compound having one negation and one disjunction.

There are sixteen Boolean functions associating the input truth values P and Q with four-digit binary outputs. These correspond to possible choices of binary logical connectives for classical logic. Different implementations of classical logic can choose different functionally complete subsets of connectives.

One approach is to choose a *minimal* set, and define other connectives by some logical form, as in the example with the material conditional above. The following are the minimal functionally complete sets of operators in classical logic whose arities do not exceed 2:

One element

$\{\uparrow\}$, $\{\downarrow\}$.

Two elements

$\{\lor, \neg\}$, $\{\land, \neg\}$, $\{\to, \neg\}$, $\{\leftarrow, \neg\}$, $\{\to, \bot\}$, $\{\leftarrow, \bot\}$, $\{\to, \nleftrightarrow\}$, $\{\leftarrow, \nleftrightarrow\}$, $\{\to, \nrightarrow\}$, $\{\to, \nleftarrow\}$, $\{\leftarrow, \nrightarrow\}$, $\{\leftarrow, \nleftarrow\}$, $\{\nrightarrow, \neg\}$, $\{\nleftarrow, \neg\}$, $\{\nrightarrow, \top\}$, $\{\nleftarrow, \top\}$, $\{\nrightarrow, \leftrightarrow\}$, $\{\nleftarrow, \leftrightarrow\}$.

Three elements

$\{\lor, \leftrightarrow, \bot\}$, $\{\lor, \leftrightarrow, \nleftrightarrow\}$, $\{\lor, \nleftrightarrow, \top\}$, $\{\land, \leftrightarrow, \bot\}$, $\{\land, \leftrightarrow, \nleftrightarrow\}$, $\{\land, \nleftrightarrow, \top\}$.

See more details about functional completeness in classical logic at Functional completeness in truth function.

Another approach is to use with equal rights connectives of a certain convenient and functionally complete, but *not minimal* set. This approach requires more propositional axioms, and each equivalence between logical forms must be either an axiom or provable as a theorem.

The situation, however, is more complicated in intuitionistic logic. Of its five connectives, $\{\land, \lor, \to, \neg, \bot\}$, only negation "$\neg$" can be reduced to other connectives (see details). Neither conjunction, disjunction, nor material conditional has an equivalent form constructed of the other four logical connectives.

Properties

Some logical connectives possess properties which may be expressed in the theorems containing the connective. Some of those properties that a logical connective may have are:

- **Associativity**: Within an expression containing two or more of the same associative connectives in a row, the order of the operations does not matter as long as the sequence of the operands is not changed.
- **Commutativity**: The operands of the connective may be swapped preserving logical equivalence to the original expression.
- **Distributivity**: A connective denoted by · distributes over another connective denoted by +, if $a \cdot (b + c) = (a \cdot b) + (a \cdot c)$ for all operands a, b, c.
- **Idempotence**: Whenever the operands of the operation are the same, the compound is logically equivalent to the operand.
- **Absorption**: A pair of connectives $\land$, $\lor$ satisfies the absorption law if {\ displaystyle a\land (a\lor b)=a} for all operands a, b.
- **Monotonicity**: If $f(a_1, ..., a_n) \le f(b_1, ..., b_n)$ for all $a_1, ..., a_n, b_1, ..., b_n \in \{0,1\}$ such that $a_1 \le b_1, a_2 \le b_2, ..., a_n \le b_n$. E.g., $\lor$, $\land$, $\top$, $\bot$.
- **Affinity**: Each variable always makes a difference in the truth-value of the operation or it never makes a difference. E.g., $\neg$, $\leftrightarrow$, {\displaystyle \nleftrightarrow }, $\top$, $\bot$.
- **Duality**: To read the truth-value assignments for the operation from top to bottom on its truth table is the same as taking the complement of reading the

table of the same or another connective from bottom to top. Without resorting to truth tables it may be formulated as $\tilde{g}(\neg a_1, ..., \neg a_n) = \neg g(a_1, ..., a_n)$. E.g., ¬.

- **Truth-preserving**: The compound all those argument are tautologies is a tautology itself. E.g., ∨, ∧, ⊤, →, ↔, ⊂. (see validity)
- **Falsehood-preserving**: The compound all those argument are contradictions is a contradiction itself. E.g., ∨, ∧, {\displaystyle \nleftrightarrow }, ⊥, ⊄, ⊅. (see validity)
- **Involutivity** (for unary connectives): $f(f(a)) = a$. E.g. negation in classical logic.

For classical and intuitionistic logic, the "=" symbol means that corresponding implications "...→..." and "...←..." for logical compounds can be both proved as theorems, and the "≤" symbol means that "...→..." for logical compounds is a consequence of corresponding "...→..." connectives for propositional variables. Some many-valued logics may have incompatible definitions of equivalence and order (entailment).

Both conjunction and disjunction are associative, commutative and idempotent in classical logic, most varieties of many-valued logic and intuitionistic logic. The same is true about distributivity of conjunction over disjunction and disjunction over conjunction, as well as for the absorption law.

In classical logic and some varieties of many-valued logic, conjunction and disjunction are dual, and negation is self-dual, the latter is also self-dual in intuitionistic logic.

Order of Precedence

As a way of reducing the number of necessary parentheses, one may introduce precedence rules: ¬ has higher precedence than ∧, ∧ higher than ∨, and ∨ higher than →. So for example, $P \vee Q \wedge \neg R \rightarrow S$ is short for $(P \vee (Q \wedge (\neg R))) \rightarrow S$.

Here is a table that shows a commonly used precedence of logical operators.

Operator	Precedence
¬	1
∧	2
∨	3
→	4
↔	5

However, not all authors use the same order; for instance, an ordering in which disjunction is lower precedence than implication or bi-implication has also been used. Sometimes precedence between conjunction and disjunction is unspecified requiring to provide it explicitly in given formula with parentheses. The order of precedence determines which connective is the "main connective" when interpreting a non-atomic formula.

Computer Science

A truth-functional approach to logical operators is implemented as logic gates in digital circuits. Practically all digital circuits (the major exception is DRAM) are built up from NAND, NOR, NOT, and transmission gates; see more details in Truth function in computer science. Logical operators over bit vectors (corresponding to finite Boolean algebras) are bitwise operations.

But not every usage of a logical connective in computer programming has a Boolean semantic. For example, lazy evaluation is sometimes implemented for $P \wedge Q$and $P \vee Q$, so these connectives are not commutative if some of expressions P, Q has side effects. Also, a conditional, which in some sense corresponds to the material conditional connective, is essentially non-Boolean because for if (P) then Q; the consequent Q is not executed if the antecedent P is false (although a compound as a whole is successful ≈ "true" in such case). This is closer to intuitionist and constructivist views on the material conditional, rather than to classical logic's ones.

OCTAHEDRON

In geometry, an **octahedron** (plural: octahedra) is a polyhedron with eight faces. A **regular** octahedron is a Platonic solid composed of eight equilateral triangles, four of which meet at each vertex.

A regular octahedron is the dual polyhedron of a cube. It is a rectifiedtetrahedron. It is a square bipyramid in any of three orthogonal orientations. It is also a triangular antiprism in any of four orientations.

An octahedron is the three-dimensional case of the more general concept of a cross polytope.

A regular octahedron is a 3-ball in the Manhattan (l_1) metric.

Regular Octahedron

Dimensions

If the edge length of a regular octahedron is a, the radius of a circumscribed sphere (one that touches the octahedron at all vertices) is

$$r_u = \frac{a}{2}\sqrt{2} \approx 0.707 \cdot a$$

and the radius of an inscribed sphere (tangent to each of the octahedron's faces) is

$$r_i = \frac{a}{6}\sqrt{6} \approx 0.408 \cdot a$$

while the midradius, which touches the middle of each edge, is

$$r_m = \frac{a}{2} = 0.5 \cdot a$$

Orthogonal Projections

The *octahedron* has four special orthogonal projections, centered, on an edge, vertex, face, and normal to a face. The second and third correspond to the B_2 and A_2 Coxeter planes.

Orthogonal projections				
Centered by	**Edge**	**Face Normal**	**Vertex**	**Face**
Image	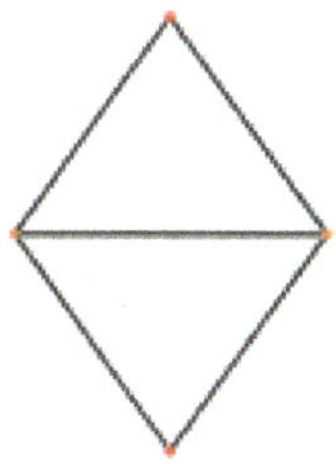	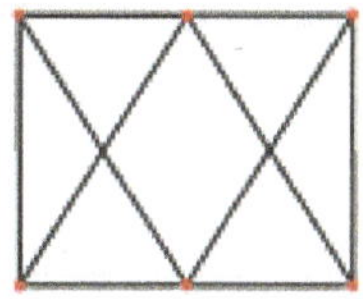	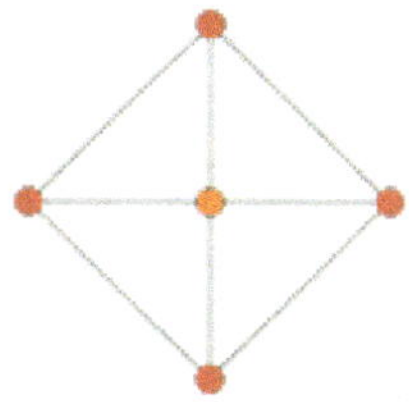	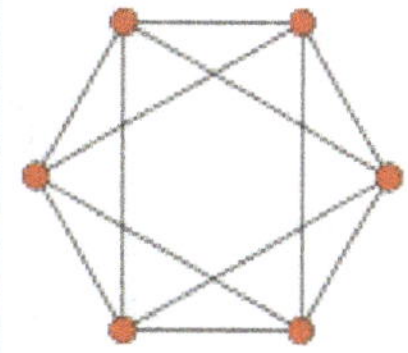
Projective symmetry				

Spherical Tiling

The octahedron can also be represented as a spherical tiling, and projected onto the plane via a stereographic projection. This projection is conformal, preserving angles but not areas or lengths. Straight lines on the sphere are projected as circular arcs on the plane.

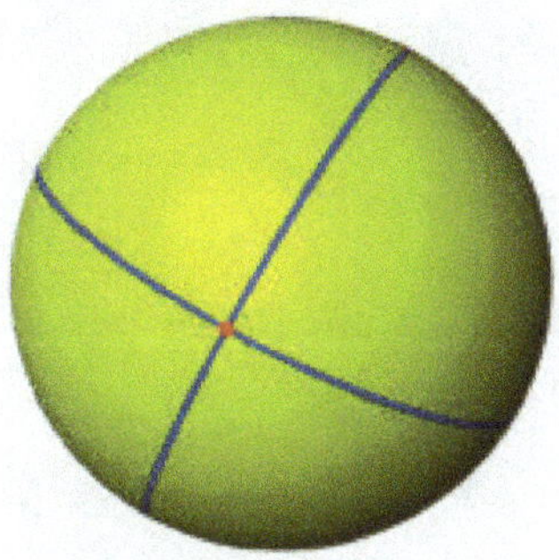	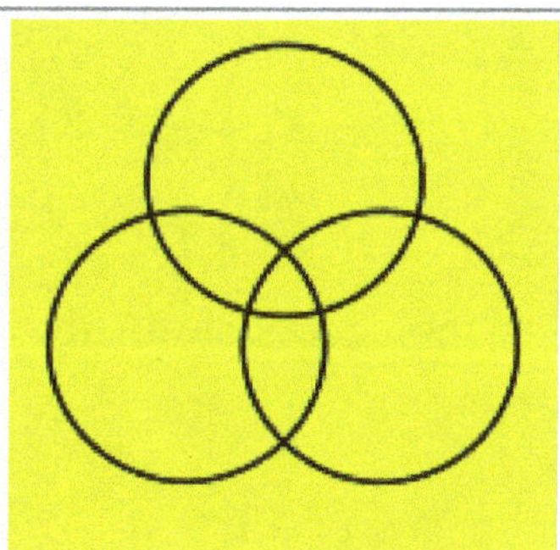
Orthographic projection	**Stereographic projection**

Cartesian Coordinates

An octahedron with edge length √2 can be placed with its center at the origin and its vertices on the coordinate axes; the Cartesian coordinates of the vertices are then

(±1, 0, 0);

(0, ±1, 0);

(0, 0, ±1).

In an *x–y–z* Cartesian coordinate system, the octahedron with center coordinates (*a*, *b*, *c*) and radius *r* is the set of all points (*x*, *y*, *z*) such that

$$|x-a|+|y-b|+|z-c|=r.$$

Area and Volume

The surface area *A* and the volume *V* of a regular octahedron of edge length *a* are:

$$A=2\sqrt{3}a^2\approx 3.464a^2$$
$$V=\frac{1}{3}\sqrt{2}a^3\approx 0.471a^3$$

Thus the volume is four times that of a regular tetrahedron with the same edge length, while the surface area is twice (because we have 8 rather than 4 triangles).

If an octahedron has been stretched so that it obeys the equation

$$\left|\frac{x}{x_m}\right|+\left|\frac{y}{y_m}\right|+\left|\frac{z}{z_m}\right|=1,$$

the formulas for the surface area and volume expand to become

$$A=4\,x_m\,y_m\,z_m\times\sqrt{\frac{1}{x_m^2}+\frac{1}{y_m^2}+\frac{1}{z_m^2}},$$
$$V=\frac{4}{3}x_m\,y_m\,z_m.$$

Additionally the inertia tensor of the stretched octahedron is

$$I=\begin{bmatrix}\frac{1}{10}m(y_m^2+z_m^2) & 0 & 0\\ 0 & \frac{1}{10}m(x_m^2+z_m^2) & 0\\ 0 & 0 & \frac{1}{10}m(x_m^2+y_m^2)\end{bmatrix}$$

These reduce to the equations for the regular octahedron when

$$x_m=y_m=z_m=a\,\frac{\sqrt{2}}{2}.$$

Geometric Relations

The interior of the compound of two dual tetrahedra is an octahedron, and this compound, called the stella octangula, is its first and only stellation. Correspondingly, a regular octahedron is the result of cutting off from a regular tetrahedron, four regular tetrahedra of half the linear size (i.e. rectifying the tetrahedron). The vertices of the octahedron lie at the midpoints of the edges of the tetrahedron, and in this sense it relates to the tetrahedron in the same way that the cuboctahedron and icosidodecahedron relate to the other Platonic solids. One can also divide the edges of an octahedron in the ratio of the golden mean to define the vertices of an icosahedron. This is done by first placing vectors along the octahedron›s edges such that each face is bounded by a cycle, then similarly partitioning each edge into the golden mean along the direction of its vector. There are five octahedra that define any given icosahedron in this fashion, and together they define a *regular compound.*

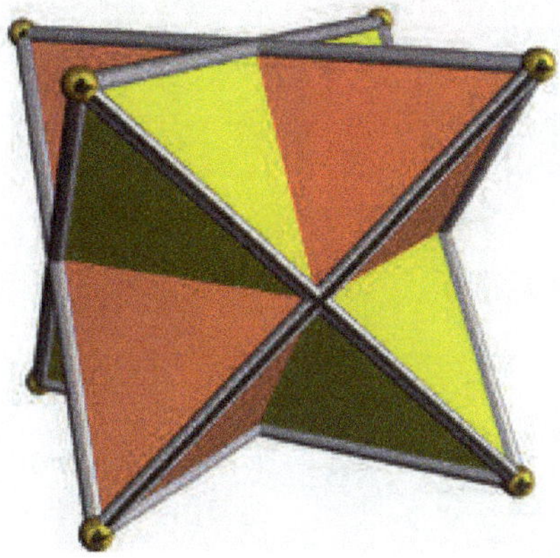

The octahedron represents the central intersection of two tetrahedra.

Octahedra and tetrahedra can be alternated to form a vertex, edge, and face-uniform tessellation of space, called the octet truss by Buckminster Fuller. This is the only such tiling save the regular tessellation of cubes, and is one of the 28 convex uniform honeycombs. Another is a tessellation of octahedra and cuboctahedra.

The octahedron is unique among the Platonic solids in having an even number of faces meeting at each vertex. Consequently, it is the only member of that group to possess mirror planes that do not pass through any of the faces.

Using the standard nomenclature for Johnson solids, an octahedron would be called a *square bipyramid.* Truncation of two opposite vertices results in a square bifrustum.

The octahedron is 4-connected, meaning that it takes the removal of four vertices to disconnect the remaining vertices. It is one of only four 4-connected simplicial well-covered polyhedra, meaning that all of the maximal independent sets of its vertices have the same size. The other three polyhedra with this property are the pentagonal dipyramid, the snub disphenoid, and an irregular polyhedron with 12 vertices and 20 triangular faces.

Uniform Colorings and Symmetry

There are 3 uniform colorings of the octahedron, named by the triangular face colors going around each vertex: 1212, 1112, 1111.

The octahedron›s symmetry group is O_h, of order 48, the three dimensional hyperoctahedral group. This group's subgroups include D_{3d} (order 12), the symmetry group of a triangular antiprism; $\mathbf{D_{4h}}$ (order 16), the symmetry group of a square bipyramid; and T_d (order 24), the symmetry group of a rectified tetrahedron. These symmetries can be emphasized by different colorings of the faces.

Name	Octahedron	Rectified tetrahedron (Tetratetrahedron)	Triangular antiprism	Square bipyramid	Rhombic fusil
Image (Face coloring)	 (1111)	 (1212)	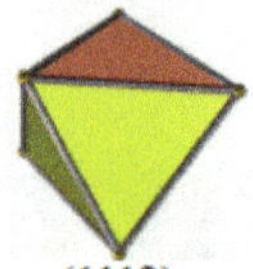 (1112)	 (1111)	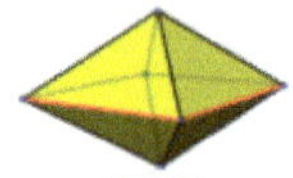 (1111)
Coxeter diagram					
Schläfli symbol	{3,4}	r{3,3}	s{2,6} sr{2,3}	ft{2,4} { } + {4}	ftr{2,2} { } + { } + { }
Wythoff symbol	4 | 3 2	2 | 4 3	2 | 6 2 | 2 3 2		
Symmetry	O_h, [4,3], (*432)	T_d, [3,3], (*332)	D_{3d}, $[2^+,6]$, (2*3) D_3, $[2,3]^+$, (322)	D_{4h}, [2,4], (*422)	D_{2h}, [2,2], (*222)
Order	48	24	12 6	16	8

Nets

It has eleven arrangements of nets.

Dual

The octahedron is the dual polyhedron to the cube.

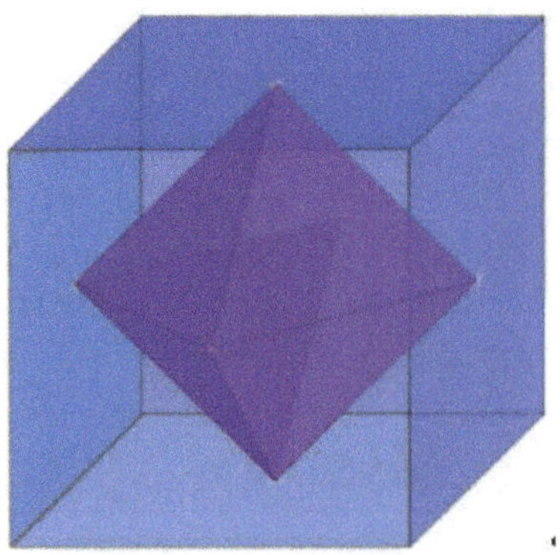

Faceting

The uniform tetrahemihexahedron is a tetrahedral symmetry faceting of the regular octahedron, sharing edge and vertex arrangement. It has four of the triangular faces, and 3 central squares.

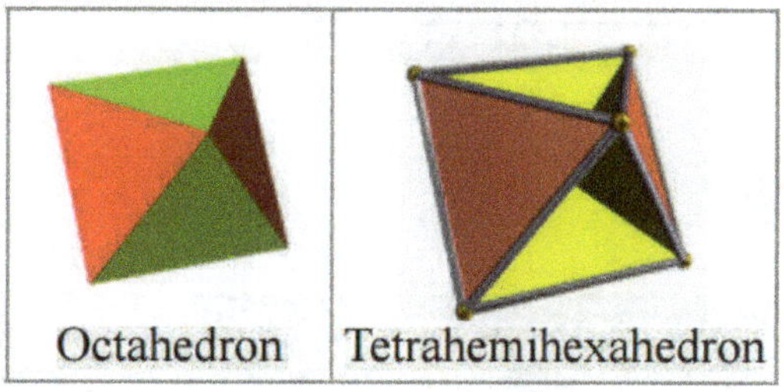

Irregular Octahedra

The following polyhedra are combinatorially equivalent to the regular polyhedron. They all have six vertices, eight triangular faces, and twelve edges that correspond one-for-one with the features of a regular octahedron.

- *Triangular antiprisms*: Two faces are equilateral, lie on parallel planes, and have a common axis of symmetry. The other six triangles are isosceles.
- Tetragonal bipyramids, in which at least one of the equatorial quadrilaterals lies on a plane. The regular octahedron is a special case in which all three quadrilaterals are planar squares.
- Schönhardt polyhedron, a nonconvex polyhedron that cannot be partitioned into tetrahedra without introducing new vertices.

Other Convex Octahedra

More generally, an octahedron can be any polyhedron with eight faces. The regular octahedron has 6 vertices and 12 edges, the minimum for an octahedron; irregular octahedra may have as many as 12 vertices and 18 edges. There are 257 topologically distinct *convex* octahedra, excluding mirror images. More specifically there are 2, 11, 42, 74, 76, 38, 14 for octahedra with 6 to 12 vertices respectively. (Two polyhedra are "topologically distinct" if they have intrinsically different arrangements of faces and vertices, such that it is impossible to distort one into the other simply by changing the lengths of edges or the angles between edges or faces.)

Some better known irregular octahedra include the following:

- Hexagonal prism: Two faces are parallel regular hexagons; six squares link corresponding pairs of hexagon edges.
- Heptagonal pyramid: One face is a heptagon (usually regular), and the remaining seven faces are triangles (usually isosceles). It is not possible for all triangular faces to be equilateral.
- Truncated tetrahedron: The four faces from the tetrahedron are truncated to become regular hexagons, and there are four more equilateral triangle faces where each tetrahedron vertex was truncated.
- Tetragonal trapezohedron: The eight faces are congruent kites.

Octahedra in the Physical World

Octahedra in Nature

Fluorite octahedron.

- Natural crystals of diamond, alum or fluorite are commonly octahedral, as the space-filling tetrahedral-octahedral honeycomb.
- The plates of kamacite alloy in octahedrite meteorites are arranged paralleling the eight faces of an octahedron.
- Many metal ions coordinate six ligands in an octahedral or distorted octahedral configuration.
- Widmanstätten patterns in nickel-iron crystals

Octahedra in art and Culture

Two identically formed rubik's snakes can approximate an octahedron.

- Especially in roleplaying games, this solid is known as a “d8”, one of the more common polyhedral dice.
- If each edge of an octahedron is replaced by a one-ohm resistor, the resistance between opposite vertices is 1/2 ohm, and that between adjacent vertices 5/12 ohm.
- Six musical notes can be arranged on the vertices of an octahedron in such a way that each edge represents a consonant dyad and each face represents a consonant triad; see hexany.

Tetrahedral Truss

A framework of repeating tetrahedrons and octahedrons was invented by Buckminster Fuller in the 1950s, known as a space frame, commonly regarded as the strongest structure for resisting cantilever stresses.

Related polyhedra

A regular octahedron can be augmented into a tetrahedron by adding 4 tetrahedra on alternated faces. Adding tetrahedra to all 8 faces creates the stellated octahedron.

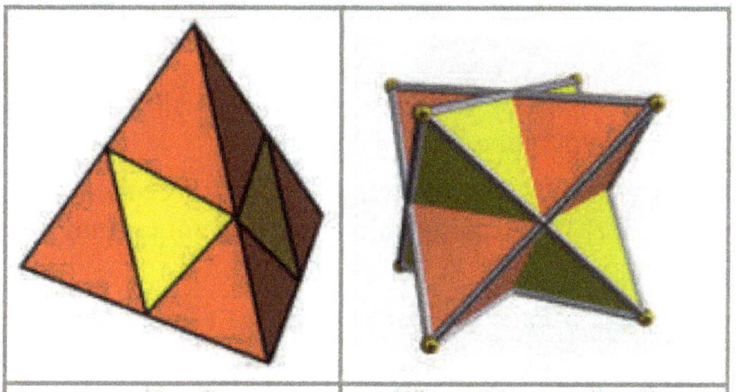

tetrahedron	stellated octahedron

The octahedron is one of a family of uniform polyhedra related to the cube.

Uniform octahedral polyhedra										
Symmetry: [4,3], (*432)							$[4,3]^+$ (432)	$[1^+,4,3] = [3,3]$ (*332)		$[3^+,4]$ (3*2)
{4,3}	t{4,3}	r{4,3} r$\{3^{1,1}\}$	t{3,4} t$\{3^{1,1}\}$	**{3,4}** $\{3^{1,1}\}$	rr{4,3} $s_2\{3,4\}$	tr{4,3}	sr{4,3}	h{4,3} {3,3}	$h_2\{4,3\}$ t{3,3}	s{3,4} s$\{3^{1,1}\}$

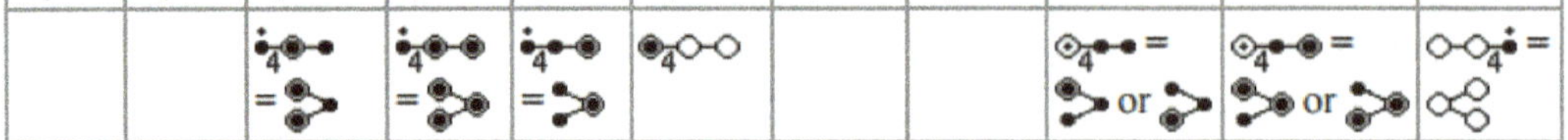										
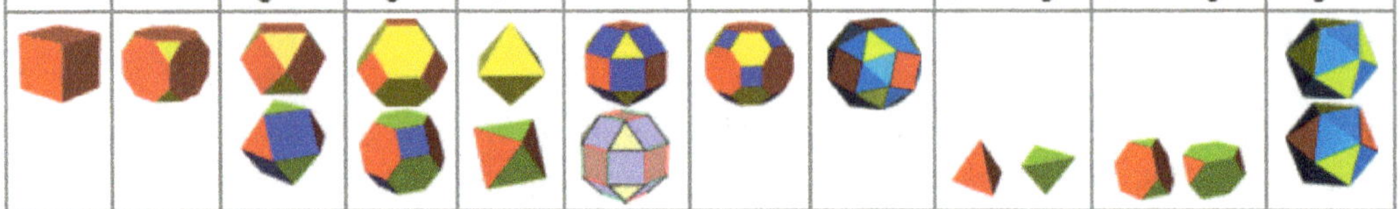										
Duals to uniform polyhedra										
$V4^3$	$V3.8^2$	$V(3.4)^2$	$V4.6^2$	$V3^4$	$V3.4^3$	V4.6.8	$V3^4.4$	$V3^3$	$V3.6^2$	$V3^5$
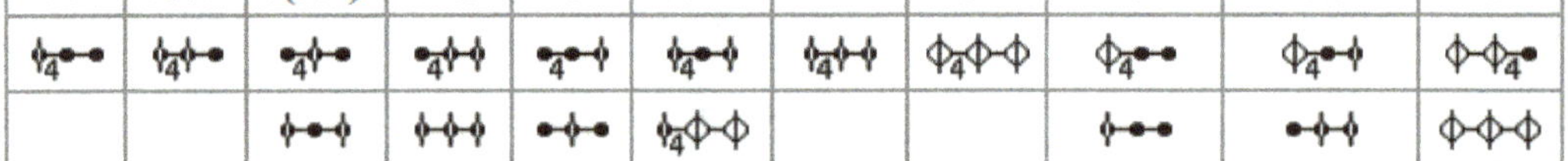										
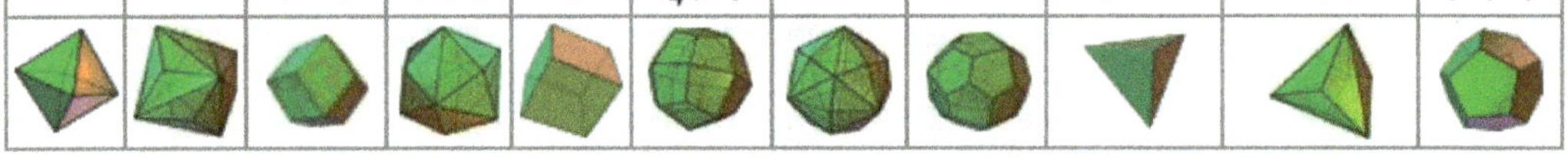										

It is also one of the simplest examples of a hypersimplex, a polytope formed by certain intersections of a hypercube with a hyperplane.

The octahedron is topologically related as a part of sequence of regular polyhedra with Schläfli symbols $\{3,n\}$, continuing into the hyperbolic plane.

*n32 symmetry mutation of regular tilings: $\{3,n\}$											
Spherical				Euclid.	Compact hyper.		Paraco.	Noncompact hyperbolic			
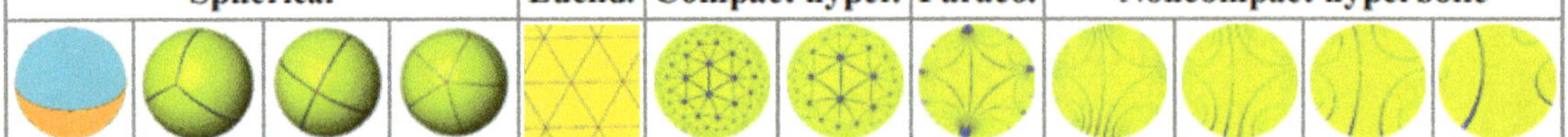											
3.3	3^3	$\mathbf{3^4}$	3^5	3^6	3^7	3^8	3^∞	3^{12i}	3^{9i}	3^{6i}	3^{3i}

Tetratetrahedron

The regular octahedron can also be considered a *rectified tetrahedron* – and can be called a *tetratetrahedron*. This can be shown by a 2-color face model. With this coloring, the octahedron has tetrahedral symmetry.

Compare this truncation sequence between a tetrahedron and its dual:

Family of uniform tetrahedral polyhedra							
Symmetry: [3,3], (*332)							$[3,3]^+$, (332)

{3,3}	t{3,3}	**r{3,3}**	t{3,3}	{3,3}	rr{3,3}	tr{3,3}	sr{3,3}
Duals to uniform polyhedra							
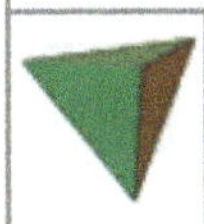		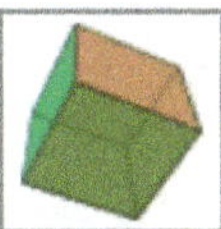	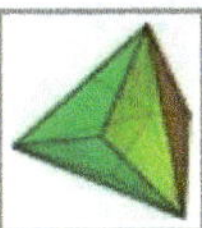			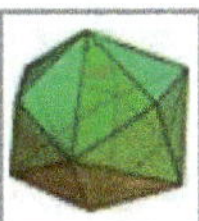	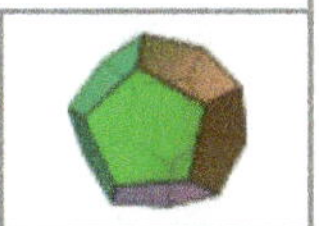
V3.3.3	V3.6.6	V3.3.3.3	V3.6.6	V3.3.3	V3.4.3.4	V4.6.6	V3.3.3.3.3

The above shapes may also be realized as slices orthogonal to the long diagonal of a tesseract. If this diagonal is oriented vertically with a height of 1, then the first five slices above occur at heights r, 3/8, 1/2, 5/8, and s, where r is any number in the range $0 < r \leq 1/4$, and s is any number in the range $3/4 \leq s < 1$.

The octahedron as a *tetratetrahedron* exists in a sequence of symmetries of quasiregular polyhedra and tilings with vertex configurations $(3.n)^2$, progressing from tilings of the sphere to the Euclidean plane and into the hyperbolic plane. With orbifold notation symmetry of $*n32$ all of these tilings are Wythoff constructions within a fundamental domain of symmetry, with generator points at the right angle corner of the domain.

$*n32$ orbifold symmetries of quasiregular tilings: $(3.n)^2$							
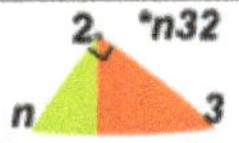	Spherical			Euclidean	Hyperbolic		
Construction	*332	*432	*532	*632	*732	*832...	*∞32
Quasiregular figures					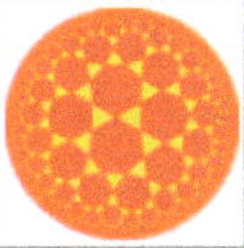		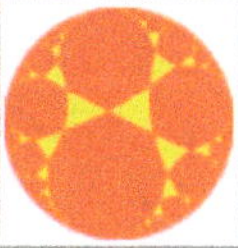
Vertex	$(3.3)^2$	$(3.4)^2$	$(3.5)^2$	$(3.6)^2$	$(3.7)^2$	$(3.8)^2$	$(3.\infty)^2$

Trigonal Antiprism

As a trigonal antiprism, the octahedron is related to the hexagonal dihedral symmetry family.

Uniform hexagonal dihedral spherical polyhedra								
Symmetry: [6,2], (*622)							$[6,2]^+$, (622)	$[6,2^+]$, (2*3)
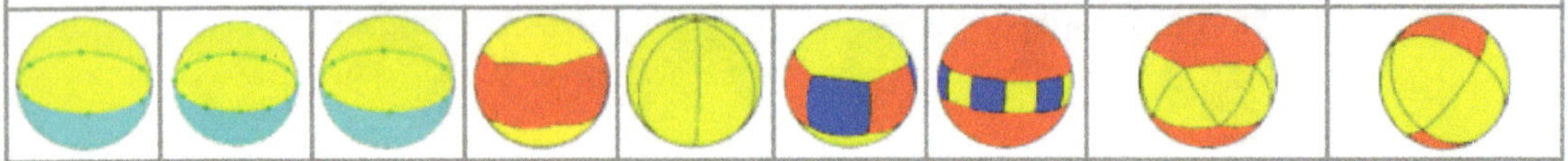								

{6,2}	t{6,2}	r{6,2}	t{2,6}	{2,6}	rr{6,2}	tr{6,2}	sr{6,2}	**s{2,6}**
Duals to uniforms								
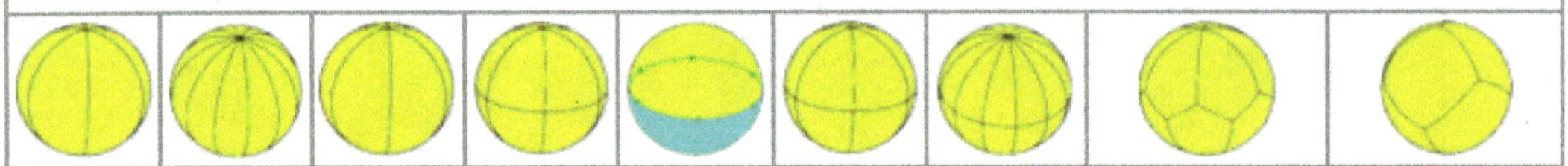								
V6^2	V12^2	V6^2	V4.4.6	V2^6	V4.4.6	V4.4.12	V3.3.3.6	V3.3.3.3

Family of uniform antiprisms *n*.3.3.3											
Polyhedron	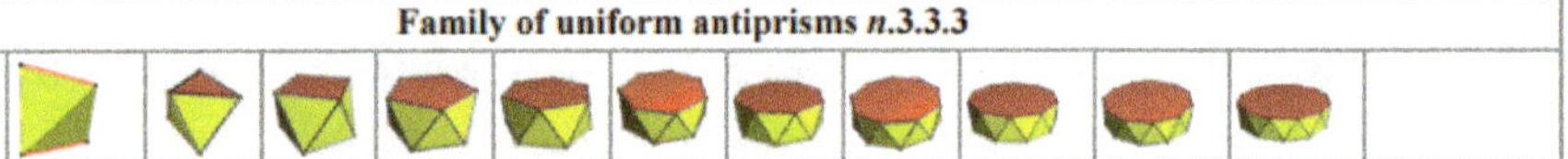										
Tiling	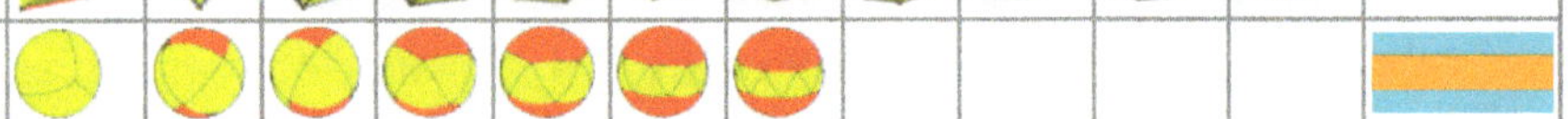										
Config.	**V2.3.3.3**	**3.3.3.3**	**4.3.3.3**	**5.3.3.3**	**6.3.3.3**	**7.3.3.3**	**8.3.3.3**	**9.3.3.3**	**10.3.3.3**	**11.3.3.3**	**12.3.3.3**	**...∞.3.3.3**

Family of bipyramids

Polyhedron	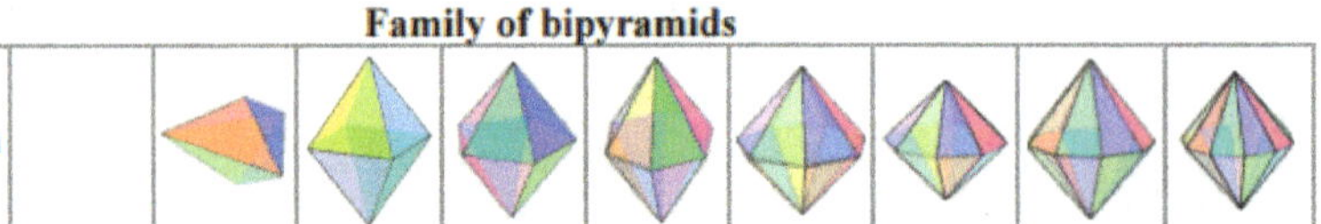								
Coxeter									
Tiling	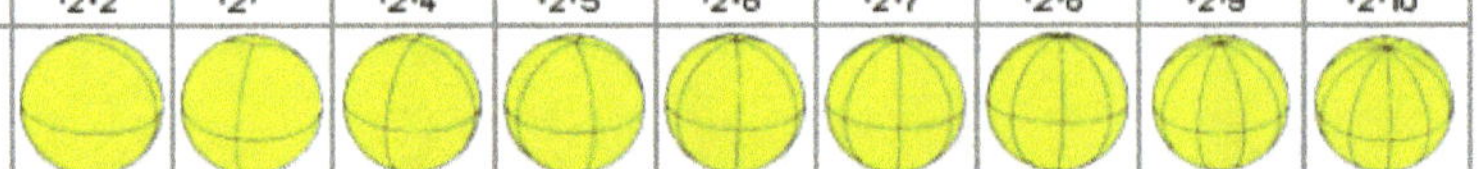								
Config.	**V2.4.4**	**V3.4.4**	**V4.4.4**	**V5.4.4**	**V6.4.4**	**V7.4.4**	**V8.4.4**	**V9.4.4**	**V10.4.4**

REFERENCES

1. Cogwheel. "What is the difference between logical and conditional /operator/". Stack Overflow. Retrieved 9 April 2015.
2. Heyting (1929) Die formalen Regeln der intuitionistischen Logik.
3. Denis Roegel (2002), Petit panorama des notations logiques du 20e siècle (see chart on page 2).
4. Russell (1908) Mathematical logic as based on the theory of types (American Journal of Mathematics 30, p222–262, also in From Frege to Gödel edited by van Heijenoort).

5. Peano (1889) Arithmetices principia, nova methodo exposita.

6. Schönfinkel (1924) *Über die* Bausteine der mathematischen Logik, translated as On the building blocks of mathematical logic in From Frege to Gödel edited by van Heijenoort.

7. Peirce (1867) On an improvement in Boole›s calculus of logic.

8. Hilbert (1917/1918) Prinzipien der Mathematik (Bernays› course notes).

9. Vax (1982) Lexique logique, Presses Universitaires de France.

10. Tarski (1940) Introduction to logic and to the methodology of deductive sciences.

11. Gentzen (1934) Untersuchungen über das logische Schließen.

12. Chazal (1996) : Éléments de logique formelle.

13. See Roegel

14. Bocheński (1959), A Précis of Mathematical Logic, passim.

15. O›Donnell, John; Hall, Cordelia; Page, Rex (2007), Discrete Mathematics Using a Computer, Springer, p. 120, ISBN 9781846285981.

16. Jackson, Daniel (2012), Software Abstractions: Logic, Language, and Analysis, MIT Press, p. 263, ISBN 9780262017152.

17. Finbow, Arthur S.; Hartnell, Bert L.; Nowakowski, Richard J.; Plummer, Michael D. (2010). "On well-covered triangulations. III". Discrete Applied Mathematics. **158** *(8): 894–912. doi:10.1016/j.dam.2009.08.002. MR 2602814.*

18. Counting polyhedra

19. http://www.uwgb.edu/dutchs/symmetry/poly8f0.htm

20. Klein, Douglas J. (2002). "Resistance-Distance Sum Rules" (PDF). Croatica Chemica Acta. **75** *(2): 633–649. Retrieved 2006-09-30.*

21. Coxeter Regular Polytopes, Third edition, (1973), Dover edition, ISBN 0-486-61480-8 (Chapter V: The Kaleidoscope, Section: 5.7 Wythoff›s construction)

22. Two Dimensional symmetry Mutations by Daniel Huson

23. Sandifer, Ed (2003). "How Euler Did It" (pdf). The Mathematical Association of America: MAA Online. Retrieved 26 October 2009.

24. Ruskey, F.; Weston, M. (June 2005). "A Survey of Venn Diagrams". The Electronic Journal of Combinatorics.

25. Baron, M.E. (May 1969). "A Note on The Historical Development of Logic Diagrams". The Mathematical Gazette. **53** *(384): 113–125.* doi:10.2307/3614533. JSTOR 3614533.

26. Henderson, D.W. (April 1963). "Venn diagrams for more than four classes". American Mathematical Monthly. **70** *(4): 424–6.* doi:10.2307/2311865. JSTOR 2311865.

27. Ruskey, Frank; Savage, Carla D.; Wagon, Stan (December 2006). "The Search for Simple Symmetric Venn Diagrams" (PDF). Notices of the AMS. **53** *(11): 1304–11.*

28. Strategies for Reading Comprehension Venn Diagrams
29. Lewis, Clarence Irving (1918). A Survey of Symbolic Logic. Berkeley: University of California Press.
30. «Euler Diagrams 2004: Brighton, UK: September 22–23". Reasoning with Diagrams project, University of Kent. 2004. Retrieved 13 August 2008.
31. John Venn (1881). Symbolic logic. Macmillan. p. 108. Retrieved 9 April 2013.
32. Edwards, A. W. F. (2004), Cogwheels of the Mind: The Story of Venn Diagrams, JHU Press, p. 65, ISBN 9780801874345.
33. Grimaldi, Ralph P. (2004). Discrete and combinatorial mathematics. Boston: Addison-Wesley. p. 143. ISBN 0-201-72634-3.
34. Johnson, D. L. (2001). "3.3 Laws". Elements of logic via numbers and sets. Springer Undergraduate Mathematics Series. Berlin: Springer-Verlag. p. 62. ISBN 3-540-76123-3.

Index